The Complete Sinclair ZX81 Basic Course

First Published in the United Kingdom
by Melbourne House

This Remastered Edition
Published in 2022 by
ACORN BOOKS
acornbooks.uk

Copyright © 1981, 2022 Subvert Limited

All rights reserved. No part of this publication may be reproduced, stored in a retrieval system, or transmitted, in any form or by any means without the prior written permission of the publisher, nor be otherwise circulated in any form of binding or cover other than that in which it is published and without a similar condition being imposed on the subsequent purchaser. Any person who does so may be liable to criminal prosecution and civil claims for damages. All trademarks remain the property of their respective owners.

This book is a page-by-page reproduction of the original 1981 edition as published by Melbourne House and Beam Software. The entirety of the book is presented with no changes, corrections nor updates to the original text, images and layout; therefore no guarantee is offered as to the accuracy of the information within.

CAN YOUR SINCLAIR ZX81 COUNT?

Some of the earlier releases of the Sinclair ZX81 have a fault in their ROM which causes some arithmetic calculations to give the wrong answers. This is caused by an 'extra' three bytes in the ROM of these ZX81s that should never have been there.

To check yours, try some of these calculations that the 'old ROM' gets wrong:

	CORRECT ANSWER	WRONG ANSWER
PRINT 0.25 ** 2	0.0625	3.142384
PRINT 4 - 0.0000000001	4	12
PRINT SQR .25	0.5	1.3591409
PRINT SQR .0625	0.25	1.847264

Some ZX81s have a hardware add-on which corrects these arithmetic errors. However, this hardware add-on does not correct another problem involved with the PAUSE function. This problem is explained further in Chapter 15.

The ultimate test is PRINT PEEK 54

If the answer is 136, then your computer is fine.

If not, we suggest you contact your Sinclair distributor, and enquire about having the 'old ROM' replaced by one of the 'new ROMs'.

INDEX

CHAPTER	TITLE	PAGE
1	Introduction	1
2	Flowcharting	11
3	Arithmetic Operators	18
4	Arithmetic Functions	22
5	String Variables	29
6	Substrings	36
7	Editing	41
8	Loops and Decisions	44
9	REM, STOP, and CONT statements	54
10	Arrays	58
11	Subroutines	68
12	Characters	72
13	Two speed computer	78
14	Output	80
15	INKEY$ & PAUSE	86
16	SAVE & LOAD programs	91
17	Top down programming	96
18	Debugging	103
19	Saving Memory	117
20	Machine Code Programs	127
21	System Variables	132
22	Beyond Graphics	138
	APPENDICES	142
	REFERENCE MANUAL	160

The Course

CHAPTER 1 # INTRODUCTION

1.1 What is a computer, and what does it do?

Simply, a computer is a tool. As much as a hammer or screwdriver is a tool designed to make actions of your hands easier, so is a computer designed to make actions of your mind easier. A screwdriver, although good at screwing and unscrewing, needs a hand to guide and turn it. So, as a tool , the computer is unable to do things on its own, and requires your mind to tell it what to do. Once told what to do - provided it is able to do so - it will do it very quickly indeed.

Unfortunately, a computer needs to be told precisely what to do, using the limited instructions that it understands. For example, consider the sentence, 'I want you to add up the numbers 1 to 10'. If you instructed your computer to do this, it will not understand. However, if you were to type:

    PRINT 1+2+3+4+5+6+7+8+9+10 (then push newline)

   - the computer would respond with :

   55

As you may have already noticed, there is no difference between what the two instructions required, but the way the instruction was phrased was very important.

   Computers are very, very fussy !!

As we said earlier, a computer will do as it is told, provided it is able to do so. The things a computer can do may be put into five general categories :

   1.   Remember data and instructions.
   2.   Interpret instructions.
   3.   Perform calculations.
   4.   Print results.
   5.   Make decisions.

At first glance, these five things do not seem like much, and can be done by almost anyone, and without the need for a fussy computer. The power of a computer is in its ability to do these things, much faster than a human.

We can illustrate this with a simple test : multiple 2 by 3 by 4 by 5 by 6 by 7 by 8 by 9 by 10.

Finished ?

Now, type into the computer the following :

  PRINT 2 * 3 * 5 * 6 * 7 * 8 * 9 * 10   (follow with newline)

The computer responded with :

3628800

Pretty quick, isn't it ?

In the above instruction, the sign * is used by the computer rather than x as the computer does not know when the x is a letter or multiplication sign. The computer also does not know when the instruction is finished, and you must tell it by pushing the NEWLINE key.

The question, 'What happens when I have more than one instruction for the computer to perform ?' must be asked. It will also have an answer.

In the next section !!

1.2 Programs, or, what to do with more than one instruction.

In this section, we will define a 'computer program' for you. We will also tell you how they may be used.

The term 'computer program' - often shortened to just 'program' is simply a series of instructions that are numbered so that the computer performs them in sequence. These numbered instructions are on their own called 'program statements'. By numbering each statement, the computer is able to keep them in order, and knows which to perform first. There are a number instructions that cannot be put in a program, and these are distinguished by their names : they are called COMMANDS, and you will be introduced to these gradually. there are not a great many of them, but they are very important, as you will see.

Type in the following program :

10 PRINT "HELLO"   (newline)

Note that the line numbers are always the first part of the statement on the extreme left-hand side. Line number up to 9999.

20 PRINT "HOW ARE YOU ? " (newline)

After the line number, comes the second part of a program statement, the statement itself. A statement is just another word for instruction, and serves to tell the computer what you want it to do. In this case, the computer will print what comes immediately after the statement.

IMPORTANT : When using the PRINT statement the computer will actually print only what appears between the inverted commas, and not the commas themselves.

In the bottom left-hand corner of the screen there is a K. This is the cursor, and is a symbol or mark that lets you know that the computer is ready for what you want to do next.

Now that you have a program in the computer you will want to use or 'run' it. This is easily done by typing the command RUN. Follow this with newline. (In future we will not tell you to type newline.)

Try it now.

The following message should have appeared :

HELLO
HOW ARE YOU?

You have just written, entered and run your first program. This simple program serves to show how easily programs may be written, and therefore, computers used.

Before we show you how to write larger and more useful programs, it is necessary to introduce you to another new word in the computer programmer's dictionary.

We will do this in the next section.

1.3 The term 'variable' and its meanings.

Although it sounds impressive, the word variable has a very simple meaning: it is a name given to a storage space in which information is kept.

Here, we introduce the LET statement. This statement is used by the computer so that it can assign (and remember) the name of a storage space, and what is in it.

In the following part of a program, the LET statement is used to assign 3 variables:

10 LET A=8
20 LET B=5
30 LET C=56

Upon seeing the LET statement, the computer knows that it must set aside part of its memory to hold a numerical value and another part for the name of that value.

In this case, the computer will set aside three parts of its memory to hold the values 8,5 and 56. It will also reserve another three places for the variables A, B and C. If you want to prove that the computer will do this, then the next statement:

40 PRINT A,B,C - may be used. Note that there are no inverted commas here. They are not needed, since the value of the variable is what you want to see, not the variable name.
N.B. Commas in the PRINT statement are used for spacing.

Enter lines 10 to 40 and type LIST. The same result may be achieved by pressing newline. The statement LIST will cause a list of the program lines to be displayed on the screen. This is a very handy statement, as it allows you see the program before you use it. If you have entered the program correctly, you may now use, or 'run' it. Another command – the RUN command – will allow you to do this.

The display should be:

8   5
56

– these being the values of A, B and C respectively.

You can combine the LET statement with simple arithmetic in the following line:

50 LET D= A + B + C

Type it in then:

60 PRINT D

– and RUN

The display should be:

8   5
56
69

If so, congratulations! You have just seem how easy it is to manipulate numbers in a computer program. In doing this, the variables A, B, C have all retained their original values, but a new variable has been introduced, that has a value that is the sum of these three. See what happens when you alter the values of one of the variables by typing in new lines with different values of the same variable.

Although we have used only single letters,

variable names may be as long as you like. You must not forget, though, that the longer the variable name, the more memory space it will take up - and therefore the less memory you will have for the program.

There are two rules for variable names that must be obeyed:

(1). The first character of the name must be a letter.
                              and
(2). Only letters or digits (numbers) may be legal parts of a variable name.

Below, you can see a brief list of legal and illegal variable names. Note that spaces are allowed, making long variable names more easily read.

LEGAL	ILLEGAL
ALFRED1	1ALFRED (does not start with a letter)
POP GROUP	POP "GROUP" (the " character is not a digit or letter)

1.4. Something NEW and CLEAR

The final part of this chapter will introduce a command and a statement to you. The first of these is the command NEW. This command is very important - although it cannot be put in a program -because when used, it will empty the memory of your computer completely. The second is the CLEAR statement. Its function is to undefine all variables that have previously been defined. The use of this command is fairly limited when very short programs are being run. However, with long programs that nearly fill the memory, it can be useful if there is a way of CLEARing the old and unused variables out of memory, so that new and usable variables may be stored.

Refer back to the program you entered, in section 1.3; we will prove the statement CLEAR does as we say.

Enter the following line:

15 CLEAR

- now LIST the program.

Isn't that great! Line 15 has been inserted between lines 10 and 20, as it should be if the program lines are to be run in numerical order. This illustrates the advantage of numbered lines, in that you can put an extra line, in wherever you like.

Almost.

All line number must be integer or whole number values, hence if the lines were numbered 1,2,3,4, etc., it would not be possible to insert a line anywhere between 1 and 4. That is why most programmers number lines in multiples of 10 or 20.

Now RUN the program.

The display should show

2/40 on the bottom left hand concer of the screen. This is a display message whenever you run your program.

8

In this case, the 2 stands for report 2 — variables not found — (due to the statement 15). The second part of this message ; 40 refers to statement 40. IE. the program halts at statement 40.

This is because the value of A was CLEARed from memory and the computer did not know what the value of A should have been.

The command NEW performs much the same task, but operates on the entire memory of the computer. It clears all the program statements as well as the variables. You use this command when you wish to empty the memory prior to loading another program.

Now test how well you have understood this chapter by attempting the following questions.

1 Which of the following are legal variables?
   a) VAR1   b) 2VAR
   c) DR. WHO   c) TOM JONES
   e) A14BK7   d) 4JOHN
(You can check your answer by typing the command
    LET (variable name)=2.
If the computer accepts the command, the variable name is legal.)

2 Write a program that will store the numbers 1 to 5 in 5 variables, called A,B,C,D and E. RUN your program, then type the command
    PRINT A+B+C+D+E
If your program is correct, the answer will be 15.

3 Add a line to your program in question 2 to print out the values of all the variables, and their sum. Use the format

    A   1
    B   2
    .
    .
  etc.

CHAPTER 2 **FLOWCHARTING**

In the previous chapter, you were shown how easy it is to write your own computer program. This is fine if you only intend to write very simple programs.

If you wish to write more complex programs, then it is often necessary to write down what you wish to do before trying to write a program. After this, you build a simple picture of how you would like the program to run. This is called flowcharting. An example of a flowchart is given below :

FLOWCHART FOR EGG FRYING

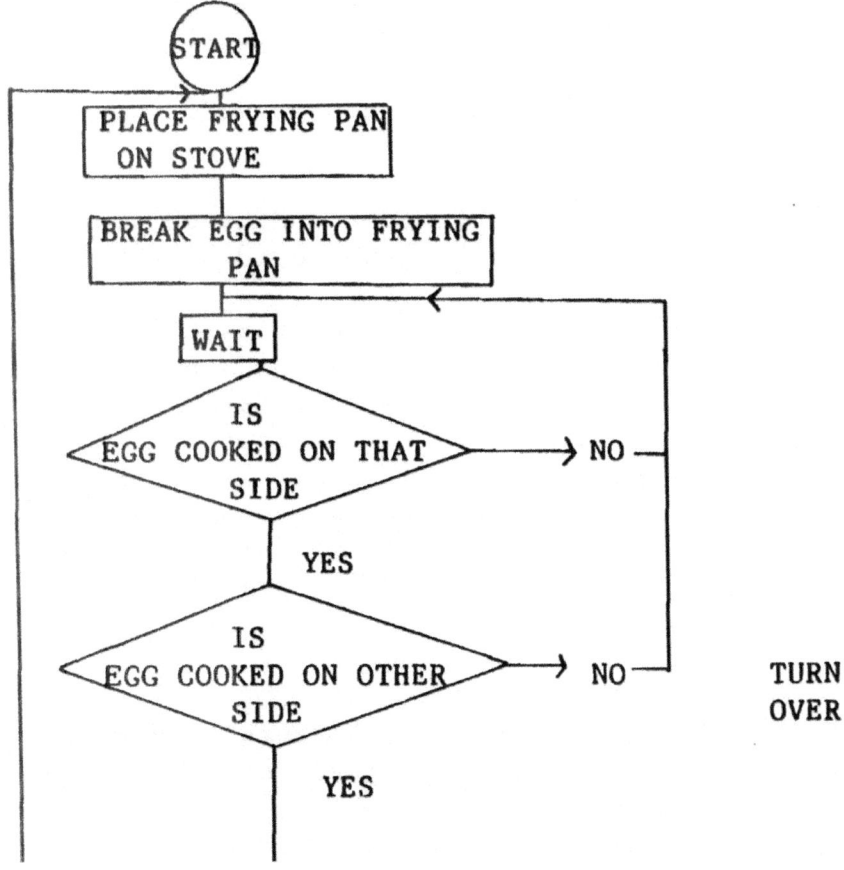

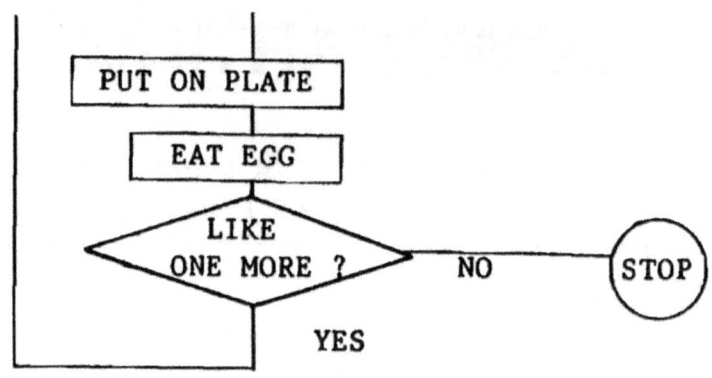

That flowchart would work well, if we wanted the computer to fry an egg for us, except that we did not have a little box containing the words "TURN ON STOVE" at the start. The egg would therefore take an extremely long time to cook. Also, when we sat down to eat the egg, we left the computer to decide what to do with the pan. If the stove was on, the pan would have been burnt.

This example illustrates the two major features of all computers :

1. They have no imagination and hence cannot assume anything (meaning we would have cold raw egg).
2. The computer does only as it is told and has no initiative (once we turned the stove on for it, the computer would promptly burn the pan).

As you saw in the example, there are a number of different shapes in a flowchart, these are to make part of it stands out, so that it may be more easily understood. We will explain to you what these shapes stand for :

⬭ This shape always contains either a START or STOP, indicating that, obviously the program starts or stops at this place. There may be only one START in each flowchart, and usually only one STOP, although in some cases, there may be more.

☐ This shape (rectangle) is usually used for all parts of the program not concerned directly with input/output, and decision making (more about these later). That is, the mathematical operations, LET statements, and so on.

These shapes must always contain a question, the answer to which may be either 'YES' or 'NO'. This is because there are times when a computer must make a decision, whilst executing a program. Since computers are only able to answer simple questions, the question must be simple.

▱ This symbol is used to signify an input/output section of a program. Output is by a PRINT statement and input by an INPUT statement. You have not yet been introduced to the INPUT statement, but its function is to halt execution while the operator (you) gives the computer a desired piece of information, or datum. The plural of datum is data, and may be either digits or characters, or in a word alphanumeric.

○ The small circle contains a number. There must always be two and only two on each complete flowchart. When writing large programs with their large flowcharts, it is sometimes not possible to put the complete flowchart on one page. Thus, these are used so

the programmer knows when there is a continuation
of a program and where the two (or more) parts
join together.

Let us now consider a situation which is more
practical in terms of using a computer: we will
write a flowchart for a computer program that
will aid in teaching arithmetic to a young child.
 The computer will give a small compliment. The
compliment is of importance, as encouragement is
always desirable when teaching. If the answer
given to the computer is incorrect, then the
computer must say so and wait for another
attempt. We will not worry about how the
computer gets the numbers or how the inputs take
place, for the moment.

Firstly, we begin with a start bubble :

And follow it with a box thus :

5c

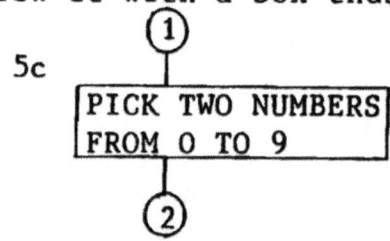

Now that we have the two numbers to be added, we
must inform the scholar-to-be what they are with
:

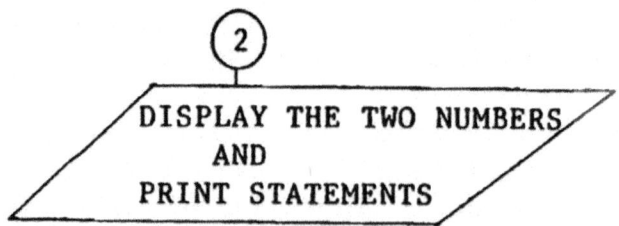

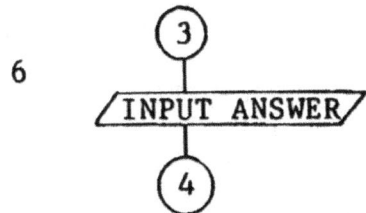

Execution of the program must now stop for the answer to be input :

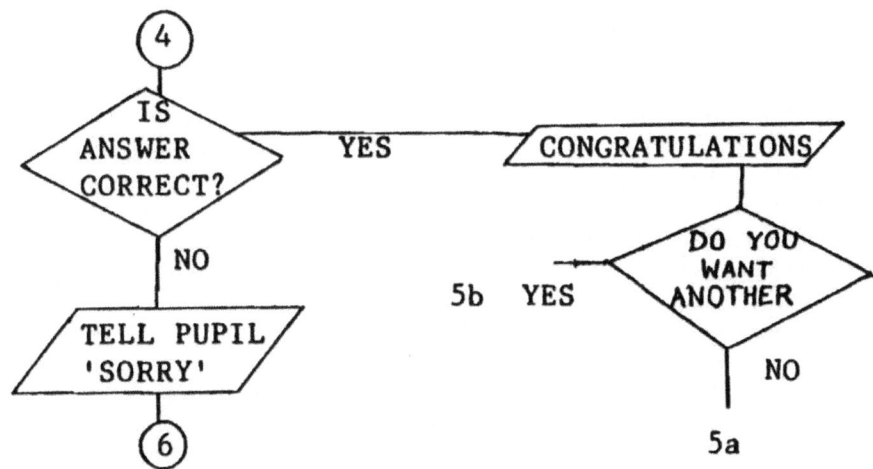

The computer must decide if the answer is correct and what to do next. If correct, the computer must display the appropriate 'congratulations' message, then ask if the pupil wishes to do another; if not correct, the appropriate 'sorry' message and invite the pupil to try again. In a flowchart, all these words would look like

The next part of the flowchart after the / DO YOU WANT ANOTHER / diamond, will not contain another input box as the question asked in the previous box is implied to have been answered by an input – either yes or no – and this requires an input from the keyboard. The next two parts of the flowchart will be :

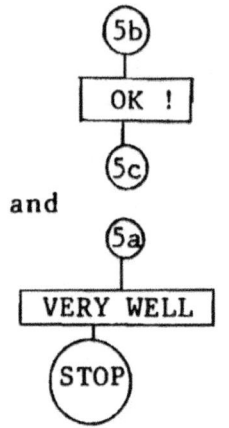

and

This signifies that execution is transferred back to the start of the program.

Referring to the case of erroreous answer, execution must be returned to the input stage.

All the pieces of the flowchart may now be pieced together and since the flowchart is for a small program, the continuation circles will not be needed :

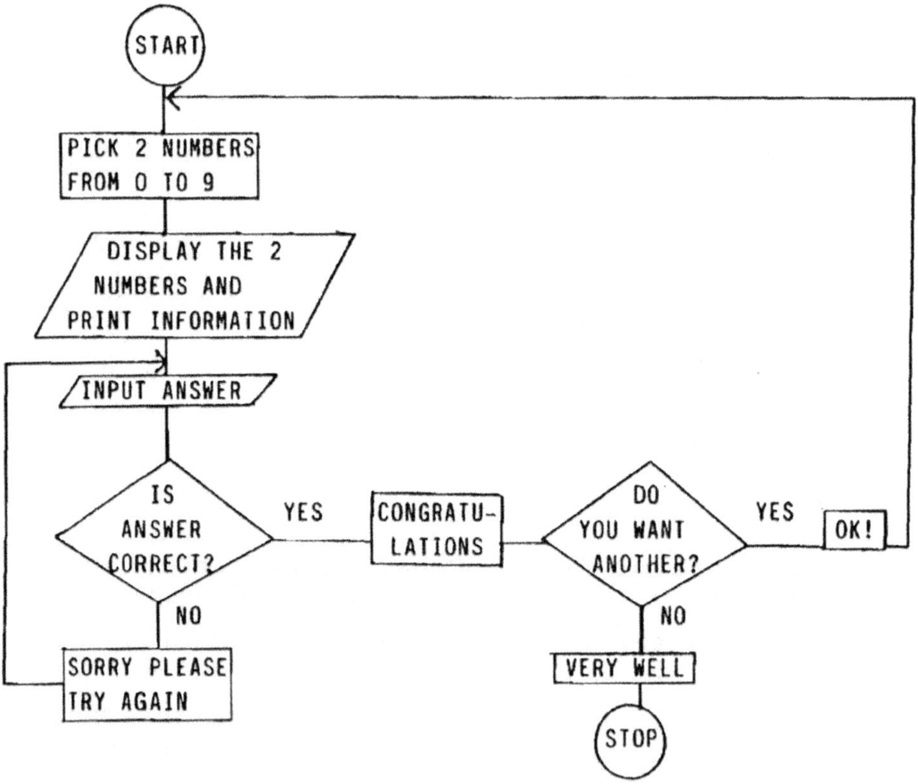

Although it seems complex now, the program will appear to be quite simple once you learn a few more statements. In fact, we will show you how to write the program later on, since you already have the flowchart written.

When writing very large programs, a useful technique is to write a very general flowchart and for each box on the flowchart, write a more detailed flowchart from which the program may be written. As an example, our egg frying flowchart may have the second box expanded to the purchase of the eggs, taking them home, placing in the fridge, and so on.

Now try to write a few of your own flowcharts. Choose tasks you know very well, like making a sandwich for instance. This may not seem very applicable to programming a computer, but you will get an idea of how to break a task into separate steps. You could also rewrite the flowchart for frying an egg, putting in the steps that were forgotten, like turning on the stove.

CHAPTER 3 ARITHMETIC OPERATIONS

In this chapter, we will show you one of the very useful facets of the ZX81 - the ability for the computer to be used as a calculator. You may not always wish to write a program if you have a short arithmetic problem, especially if you need the answer quickly.

There are five basic arithmetic operations that your ZX81 can perform:

	OPERATION	SYMBOL
1.	ADDITION	+
2.	SUBTRACTION	-
3.	MULTIPICATION	*
4.	DIVISION	/
5.	POWER	**

- as you saw, several of the ZX81 symbols are not the normal symbol and there are some good reasons for this :

(1) The first one stems from the keyboard, but when added to the lack of imagination (which computers suffer from), means that the following distinction cannot be made:

A5 x 5   and   A5 X 5   (as a variable)

Originally, computer keyboard only had capital or upper case letters, and the computer could not distinguish between a multiplication sign and an X. Leaving spaces does not help, as a computer does not recognize a space, although it will notice when one occurs, and remember it's place.

(2) Also, it is not normally possible to type in subscript (and superscript) characters, so a way around this had to be found to tell the computer when to perform, say, 5^2 .

(3) The division sign – called slash – is not radical departure from the norm in that it has long been used in fractions. It is important to remember that when the slash appears, all numbers and variables to the right, will be divided, i.e.:

        8 + 2 / 6 * 2   will be performed as :

        8 + (2 / 6 * 2),   even though you may have wanted :

        (8 + 2 / 6) * 2

In longer arithmetic operations (this is what progammers call "sums"). You may use brackets so that the line is easier to read. However, you need not do so, as the computer uses a priority system of deciding which operation to perform.

Enough theory for now, let's prove (and practice) what you have just learnt :

Type in the following lines and see if you can predict the answer before hitting NEWLINE   :

```
PRINT 2 + 2
PRINT 6 - 3
PRINT 4 * 7
PRINT 9 / 3
PRINT 4 * 4 * 2
PRINT 9 + 5 - 4 / 8
PRINT 3 ** 3 / 3 * 2
```

You can tell your computer to perform even longer operations, if you please and it will perform them with ease.

As we said in chapter 1, a computer will remember data. If, in a program you were to mix variables with your numbers, the computer would remember the values of the variables, and perform the desired operation. This feature will be pretty handy, as you will see. For now, let's see if what we're just said is correct with this short program (firstly use the command NEW to empty the memory) :

10 LET   A=10
20 LET   B=5
30 PRINT A+B*5

- now RUN it and see.

Each of the operations has a priority level attached to it, on a scale up to 16. The higher priority level, determines which operation is performed first.

Operations with equal priority are performed in order from left to right. We have listed the operations with their appropriate priority levels below :

OPERATION	PRIORITY LEVEL
Slicing	12
PI and RND	11
**	10
- (negative)	9
* and /	8
+ and -	6

The - operation can be used to experess a negative number, but the + operation may never be

used to express a positive number. The computer
will assume a number to be positive unless told
otherwise.

The two operations PI and RND are two arithmetic
functions which will describe in chapter 4.

Chapter 7 will introduce you to slicing.

QUESTIONS

1 Check that you understand the priority scheme
by typing the following commands into your ZX81.
Before pressing NEWLINE, try to predict what the
answer will be.
    a) PRINT 2+3*4
    b) PRINT 2*5**2
    c) PRINT 7+2*2**2
    d) PRINT 4--2**2

2 Add parentheses () to the following command so
that the computer gives the answer 15
    PRINT 20-7-4-2

CHAPTER 4 # ARITHMETIC FUNCTIONS

This chapter will show you what the functions many functions of the ZX81 are, and you can do with them. Some are of particular use in programs, while others are of a more specialized use. These functions are essentially the same as those found on calculators that are termed 'scientific'.

Question: What is a 'function'
Answer: A function is a rule used to obtain results.

All the functions are built into your ZX81 and are accessed (this is what computer programmers say instead of 'used') in the following manner:

1. Press FUNCTION – by SHIFT NEWLINE – and observe that the cursor changes from K to F.

2. Press the required function key. After the function appears on the screen, the cursor will revert to L.

There are 14 functions in your ZX81, and they are listed below, with the ZX81 symbol:

FUNCTION	ZX81 SYMBOL
Square root	SQR
Sign	SGN
Absolute value	ABS
Sine	SIN
Cosine	COS
Tangent	TAN
Arcsine	ASN
Arcosine	ACS

Arctangent	ATN
Natural logarithm	LN
Exponential	EXP
Integer	INT
Pi (π)	PI
Random number	RND

In order to use these functions with the exception of the last two, an argument must be used. An argument is a number or variable after the function. The only limitation on the argument is that it must mot be outside the range of values usable by the particular function.

In case you are not familiar with the functions, we will explain them to you as we go along.

1. SQR : this function will return the SQuare Root of the argument. This function can only accept positive numbers. Try the following line:

        PRINT SQR 16

Display : 4

In a program, you could use SQR like so:

    10 LET A=36
    20 PRINT "THE SQUARE ROOT OF ";A;" = ";SRQ A

2. SGN: This function can return one of several answers when used:

      a. −1 is returned if the value is less than zero.
      b. 0 if the argument is equal to zero.
      c. 1 if the argument is greater than zero.

Try the following lines:

        PRINT SGN 4

Display : 1

        PRINT SGN -20

Display : -1

        PRINT SGN 0

Display : 0

In a program this could appear as :

        10 LET A=4
        20 LET B=-20
        30 LET C=0
        40 PRINT SGN A,SGN B,SGN C

Display : 1   -1
        : 0

This function can be particularly useful if you wish execution of a program to alter depending upon the value of a variable, for example.

3. ABS : This function will return the ABSolute value (or modulus) of the argument. This means the sign - either positive or negative - of the argument is made positive. Prove this with the following examples :

        PRINT ABS 2

Display : 2

        PRINT ABS -1234

Display : 1234

4. INT : This function will cause the computer to disregard the part of an argument to the right of the decimal point (if the argument is numeric). If the argument is a variable, only the whole number part will be recognized. In other words, it will return the INTeger part of the argument. Note that this function will not round off the argument, it truncates (that is, cuts off the decimal part)

        PRINT INT 1.5694

Display : 1

        PRINT INT 46.9654

Display : 46

        PRINT INT 9.2345

Display : 9

One way of using this function is to control the number of decimal places in a number. Suppose you wish to obtain only two decimal places of the number 1.36782.

The steps to take are :

    a. Multiply the number by 100 to give 136.782

    b. Take the integer part by applying INT.
Display : 136

    c. Divide the result in step (b) by 100 to give the final result : 1.36

If you desired three decimal places, then you must use 1000 in steps (a) and (c) in place of 100.

Also, if you wished to round off the number then
before step (b), you would have to add 0.5 to the
number.

5. PI : This function needs no argument and it
returns only one value:- that of pi to ten
decimal places (although only eight will be
displayed) :

    PRINT PI

Display : 3.14159265

Whenever you need to use pi, then just use this
function.

6. RND : This is the second function that
requires no argument. Use of this function will
generate random numbers between 0 and 1, where
zero is inclusive but not 1.

RND does not actually generate random numbers but
follows a fixed sequence of 65536 numbers that
are so jumbled as to appear random.

The statement RAND is usually used in conjunction
with RND. Its function is to control the
randomness of RND. Try :

    10 RAND 1
    20 PRINT RND

Display : 0.00227355

When you use the above program, the same sequence
of random numbers will always appear on the
screen. The number displayed was the starting
number. Note that RAND does not produce a random
sequence.

In order to obtain a random sequence, try the

following program:

```
10 RAND 0
20 PRINT RND
```

After you have run the program a few times, you should realize the results are quite different. This is because RAND 0 starts RND by judging how long the television has been on. Therefore, this should be random.

7 - 14 : SIN, COS, TAN, ASN, ACS, ATN, are all trigonometry functions, and operate only in radians, not degrees. These functions perform the same operations on the argument as your calculator.

LN, EXP are mathematical functions with their own special uses.

If you have ben taught trigonometry and mathematics in general, you will probably have already thought of uses for some or all of thes functions. If not, do not worry, as you will still be abe to program your ZX81 successfully.

We will show you the general format of the above mathematical functions, and if you wish, you can try them on your ZX81 :

```
PRINT SIN 0.5
PRINT COS 0.5
PRINT TAN 0.5
PRINT ASN 0.5
PRINT ACS 0.5
PRINT ATN 0.5
PRINT EXP 2
PRINT LN 2
```

Display :   0.47942554
        :   0.87758256

```
: 0.54630249
: 0.52359878
: 1.0471976
: 0.46364761
: 7.3890561
: 0.69314718
```

Since the trigonometric functions work only in radians, you can convert degrees to radians by dividing the degree value by 180 then multiplying by pi, like so:

```
PRINT TAN(45/180*PI)
```

— Where 45 is the value in degrees. The display should read :
1

QUESTIONS

1. Use the RND function to obtain an integer number between 1 and 6. You will also need to use the INT function, and several arithmetic operators.
This statement will be used in game programs that require a die to be thrown.

2. The length of the circumference of a circle is 2 * Pi * R (R is the length of the radius). Write a command that will print the length of the circumference of a circle with a 4cm radius.

CHAPTER 5 **STRINGS**

5.1   Strings introduction

The previous chapter let you in one of the ZX81's secrets. That is, you can use ZX81 as a calculator.

Now, you will learn another secret of this machine; the computer can manipulate WORDS, as it manipulates variables!

In computing areas, these WORDS are known as strings. A string is easily recognised by the $ (dollar) sign.

The example below will introduce you more to the concept : type in the following,

       10 LET A$="I AM YOUR ZX81, AND I DO WHAT I AM TOLD."
       20 PRINT A$

As you pressed the key NEWLINE, the sentence was:-

I AM YOUR ZX81, AND I DO WHAT I AM TOLD.

- will appear on the screen. One thing you should notice is that the printed message is enclosed by quotation marks. ZX81 has printed the string in the same form as quoted.

If the string is not enclosed by quotation marks, your ZX81 will treat it as a variable; which has been mentioned before.

5.2   String variables:

Section 5.1 has allowed you to find out :

1.   What a string is?
2.   How to instruct your ZX81 to print out strings.

Now, you will learn all the ways to manipulate strings.

Chapter 1 has introduced you to variables, now you will learn more about them.

Variable names can be used to assign numbers, but presently, you will soon realize that you can assign strings to all variables as well. The previous example illustrated this to you.

A LET statement allows you to assign both strings and numbers. You must learn the difference between a string variable and a numeric variable.

For string variables, each variable name consists of two parts :

1.   a single letter (- name)
2.   a "$" sign after the first part.

We will now give you the two simple rules for naming string variables :

1. Each string variable only contain a single letter.

2. Each string variable must have a "$" sign after the name. (This doesn't apply to numeric variables)

NOTE : You mustn't forget the two rules, as if your string and numeric variables are not correctly named, then your ZX81 will become confused.

5.3   Operations with string variables

You know now that your ZX81 can perform arithmetic operations for any numbers.

Your ZX81 can also add string variables together. This operation is usually known as concatenation. This means that separate string variables will be chained together as one string variable.

An example will explain the above point more clearly.

    PRINT "BIRDS" + "△CAN FLY !"

- the display should appears as below :

BIRDS CAN FLY !

                OR :

    PRINT "A" + "B"
- you should obtain the following result :-

AB

This operation can also be used in a program. We will combine it with some strings.

    10   LET A$ = "ALL ANIMALS THAT HAVE FEATHERS"
    20   LET B$ = "△CAN FLY."
    30   PRINT A$ + B$

After you have RUN the program, the following will appear on the screen.

ALL ANIMALS THAT HAVE FEATHERS CAN FLY.

You should have a clear idea of this operation by now. Your ZX81 can only perform addition with string variables. Other operations such as subtraction, multipication, division or raise to the powers are out of your ZX81's ability.

5.4   The funstions LEN, VAL and STR$

This section will introduce you to three more "words" of the ZX81's vocabulary. These are string functions and will allow you to use strings with numbers and visa versa.

The first of these is the function LEN. This function will allow you to see how long a string is. Consider the following line:

PRINT LEN" HOW LONG IS THIS STRING?"

- the result of this would be to display the number 25. If you count the number of characters - and remember in this case a space counts as a character - you will see that the characters number 25. This can be of use if you write a program that requires a decision to be made (by the computer) during program execution, or as a check on the length of an answer. If the answer is too long, the message,"YOU TALK TOO MUCH" can be displayed, for example.

VAL is a very useful function as it allows you to perform some very interesting tricks with strings. We had better take you through them one at a time:

(1)   FOr any given string, VAL will return a numeric VALue of that string. Therefore, if your string consists of numbers, you can perform arithmetic functions on it, like so :

```
        LET A$="1234"
        PRINT VAL A$*2
```

Display : 2468

Your ZX81 takes the numeric VALue of the string A$, then multiplies by 2 to obtain the above result.

The prime rule when using VAL, is that the string must be digits, either positive or negative. The value of that string will be returned. To prove this, see what happens when you type in :

```
        10  LET B$="A1234"
        20  PRINT VAL B$ * 2
```

You should have an error message 2/20 on the bottom left hand corner of the screen; implies undefined variable at Line 20.

Also, Val must be the first function of the line.

(2) You can convert the string variable into a numeric variable like so :

```
        10 LET C$="1234"
        20 LET C=VAL C$
        30 PRINT C$,C
```

Display : 1234   1234

The same rules as in (1) also apply here.

(3) It is possible to eVALuate string as an arithmetic expression:-

```
        10 LET D=16
        20 LET E=9
        30 PRINT VAL"0.25+D*2+E/3"
```

Display : 35.25

Again as in (1) the same rules must apply.

As you can see the VAL function can be a very, very useful function.

Wouldn't it be nice, though if we could convert numeric variables into string variables?

This is precisely what STR$ does!

We can illustrate this most simply by:

```
10 LET D=6789
20 PRINT STR$ D
```

Display : 6789

The STR$ will also cause numeric values to be displayed as if they were strings.

For example :

```
PRINT STR$ 123
```

Display: 123

QUESTIONS

1. Which of the following are legal names for string variables?
    a) NAME$ b) A
    c) X$ d) 1$
You can check your answers by typing the command
   LET (variable name)="STRING"
The computer will only accept legal names.

2. Type in the following commands, and before

typing NEWLINE try to predict the result.
- a) PRINT VAL "SQR 4"
- b) PRINT LEN "SQR 4"
- c) PRINT STR$ SQR 4
- d) PRINT VAL STR$ LEN SQR 4

CHAPTER 6 **SUBSTRINGS**

In chapter 5, we introduced you to strings and string functions. In this chapter we will deal with pieces of strings, termed "substrings". At this stage, it is necessary to introduce you to another statement in your ZX81's vocabulary. This statement allows you to enter a datum into the computer while program execution is halted temporarily. The statement may only appear in a program, so you cannot use it as you would a function or PRINT; it is the input statement. Accompanying the INPUT must be a variable – either numeric or string – the value of which is the data that has been entered. The program below will illustrate its use:

```
10 PRINT "TELL ME A NUMBER PLEASE :"
20 INPUT A
30 PRINT " THE NUMBER IS ";A
40 CLS
50 PRINT "PLEASE GIVE ME A WORD :"
60 INPUT A$
70 PRINT " THE WORD IS ";A$
```

As you run the program the K cursor will change to an L cursor at the bottom of the screen. Also note that when a string is to be INPUT, the L is enclosed by inverted commas. The reason for the change of cursor, is so you will know the computer is waiting for you to supply it with information. Please note that your computer displays great patience and waits until you give a number and type NEWLINE before continuing program execution.

If the computer is expecting a number, you have

to input a number , otherwise an error message will be sent to you. The same happens for string variables.

Another new statement CLS is introduced to you in line 40. This statement will CLear the Screen - but nothing else. Any PRINT statement after CLS will be displayed as first line on the screen - top of screen.

As we said earlier this chapter is about substrings, so we will leave the INPUT statement for a while.

Consider the string "ABCDE". Since substrings are parts of a string - in correct sequence - then the following must qualify as substrings :

                ABC
                CDE
                BCDE

How do we get substrings? The answer is simple: just slice off the appropriate pieces of the string. We do this like so:

        "string expression"(start TO finish)

for example :

        "ABCDEFGH"(4 TO 6)

- would return the fourth to sixth characters (inclusive) of the string as the substring "DEF".

There are four things to be wary of when you wish to generate substrings :

(1). If you omit the starting position, then your ZX81 will interpret this as meaning you wish to be the first character of the string. That is,

        "ABCDER"(TO 5)   will return
        "ABCDE"   as the
substring

(2). Similarly, with the finishing position omitted, the interpretation will be to start at the desired position, but continue until the last character of the string is reached. Thus :

        "ABCDERGHIJKL"(3 TO)
 - will be returned as :
        "CDEFGHIJKL"

(3). Predictably, with both starting and finishing positions omitted, the whole of the string will be returned as the substring. Hence with the situation :

        "ABCD"(TO)
 - you will get
        "ABCD"   as the substring

(4). Finally, if only one number is enclosed in the brackets, then a single character will returned as the substring, in the position specified by that number :

        "ABC"(2)
 - will return the character B as the substring.

The above requirements can be met fairly easily and at times can be useful, but there are three ways you can confuse your ZX81 when dealing with substrings. We will tell you about these so that you can avoid them. Then should you accidently confuse your ZX81 with substrings, then you will be able to find out where you went wrong.

(1). If the starting position is greater than the finishing position, then your substring returned will be empty or valueless :
for example :
    "ABCDEFG"(8 TO 7)   - will see
    "" returned as the substring.

(2). If the finishing position is greater than the actual string length, then an error code will be displayed :
for example :
    "ABC"(2 TO 4), will cause an error message to be displayed as the string is only three characters long. Your ZX81 must stop looking for the fourtth character (as it does not exist) and you are told of this action accordingly.

(3). The last trap to avoid is to ensure the starting and finishing positions you have entered are both positive.

Now, type in the program and RUN it :-

```
    10 PRINT "INPUT A 8 CHARACTERS STRING ?"
    20 INPUT A$
    25 IF LEN A$ < 8 THEN GOTO 10
    30 PRINT "GUESS THE SUBSTRING(5 TO 8) !"
    40 PRINT "STRING IS ";A$;"ANSWER IS ";A$(5 TO 8)
    50 PRINT "SUBSTRING(4 TO 6)! "
    60 PRINT "STRING IS ";A$;"ANSWER IS ";A$(4 TO 6)
```

After you RUN the program few times ; you may like to go on to next chapter !

QUESTIONS

1. If A$="MNOPQRSTU", what would the following commands print? Check your answer by typing in
     LET A$="MNOPQRSTU"
and then typing in each command.
    a) PRINT A$(3 TO 5)
    b) PRINT A$(TO)
    c) PRINT A$(5 TO 10)
    d) PRINT A$(5 TO 9)+A$(4 TO 7)
    e) PRINT A$(7 TO 3)

2. In Chapter 4, question 2, you wrote a program statement that calculated the length of the circumference of a circle. Using this statement (with minor modifications) write a program that will ask for the length of the radius, and then calculate the circumference.

3. Write a program that asks for your name, and then says HELLO N, using your name.

CHAPTER 7 **EDITING**

In this chapter, we will show you what can be done when you have made a mistake entering a line. When a program line needs to changed after pressing NEWLINE, it is termed 'editing'. There are a number of editing functions in your ZX81, and we shall deal with all of them.

We have deliberately left this chapter until now so you could have practice learning where the commonly used statements can be found on your ZX81. Also, it gave you practice at typing in lines correctly.

When you have entered program lines as in previous chapters, you will hopefully have noticed an unusual symbol - like a letter V on its side - displayed between the line number and statement. This symbol is called a cursor and shows you which line is available for editing.

Before we tell you how to edit programs, we will tell you about each of the editing commands available to you.

(1). RUBOUT: on the keyboard there is a key marked RUBOUT on the top right hand corner. If you realize your mistake immediately after you make it, then by using the RUBOUT key you will be able to erase that character. You can gain access to this key by pushing SHIFT - on the lower left-hand side of the keyboard and then 0.

As you type in your line, the cursor K will move along it. When you use RUBOUT the cursor moves backwards in accord. On this

41

example use   RUBOUT   to erase C,D and E

    10 PRINT A,B,C,D,E

(2). LEFT AND RIGHT CURSOR Controls : If you reach the end of a line only to find a character in the middle has been typed in wrongly, you may move the cursor K backwards along it to correct the mistake. This is achieved by using the left ⇦ and right ⇨ arrows for cursor movement. The ⇦ will allow you move the cursor backward along the line to the mistake. You then type over the mistake and use the ⇨ to move to the end of the line. If you are satisfied the line is as you want it, you can then press NEWLINE
   Try it on this example :

    10   PRINT "CORRECT MISTALES EASILY"
                      #
                  this character should be a K

(3). DOWN arrow : As you enter each line into your ZX81, you will have seen it move from the bottom of the screen to the top. When you wish to EDIT a line, it must be moved from the top to the bottom. To do this the ⇧ and ⇩ symbols are used. The effect of using these is to move the program cursor (our V on its side) up or down the program listing. When you have the cursor on the right line, you then press   SHIFT   EDIT . The effect of this is to bring that line to the bottom of the screen so that editing can now take place.

for example :

        # program cursor
      10▶PRINT "HELLO"
-   If you press   SHIFT   EDIT   , a copy of line 10 will appear on the bottom of the screen.

(4). Line Deletion : If you decide a line is no longer needed, then that line may be completely removed from the program by typing the line number then   NEWLINE .

Try it with :

   10   PRINT " I DO NOT WANT THIS LINE"

then follow it with :   10   NEWLINE
You will see line 10 at the top of the screen disappear.

(5). Line Replacement : If you wish to replace a line, then proceed as for 4, but instead of presssing NEWLINE   type in what you wish the new statement to be and then   NEWLINE   . You will see the new line placed at the top of the screen. The following example will illustrate this:

   10 PRINT "I DO NOT THINK THIS LINE WILL DO "

   then

   10 PRINT "THIS LINE IS MUCH BETTER "

With these facilities it is possible to quickly correct any mistakes in your program entry.

CHAPTER 8 LOOPS AND DECISIONS

Although the statements we have introduced you to so far have been quite useful, the truly powerful statements havé been missing. This is because if they were introduced at the start, their significance may have not have been too apparent.

8.1 Loops without end

Let us suppose you had a number of items of data — costs for the running of a small business — and you wished to add them together and get a progressive total as you went from one item to the next. One way to do this would be to write a program that adds two numbers together and displays the answer. If you did this, then it would be necessary to type in two numbers to run the program: the next item and the progressive total.

An easier way would be to have a program that would wait for each piece of information, do what you want and return to the start for the next datum. We can do this with an unconditional GOTO statement. The word unconditional in this case refers to the fact that regardless of what has happened before in the program, execution will be transfered to the line number. This is opposed to conditional, which means that execution may be passed to the line number, depending on the value of a designated variable. More of this later.

Below is a program that shows the use of the unconditional GOTO to tackle the problem mentioned earlier. Before entering it, clear the ZX81's memory (with the command NEW).

```
10 PRINT "ENTER YOUR NUMBER WHEN THE CURSOR
   CHANGES TO  L ."
20 LET B = 0
30 INPUT A
40 LET B=B+A
50 PRINT A,B
60 GOTO 30
```

Let us draw your attention to line 40 for a minute. If you have not programmed a computer before, this line may appear to be one that will completely baffle the computer. Such is not the case! Instead of not being able to decide what value to assign to B, the computer reads this line as :

> Set a new value for varialbe B that will be equal to the old value of variable B plus the current value of variable A.

Now you know how the computer reads line 40, the whole program should make more sense.

Something however, is missing ....
Question : How does the computer know when to stop?
Answer   : It doesn't!

A new (and very important) command to learn about is BREAK. This command, when issued BREAKs into a program − regardless of where in the execution the computer is − and stops execution that point.
 Without this command the loop formed by lines 60 and 20 would continue forever. This is termed 'endless loop'.

Before RUNning the program, make up some data to serve as costs to be input. Then type RUN. When you have run out of data, use the command BREAK to halt execution. Try drawing a flowchart for the program as well.

8.2 Loops that do end

There are two ways you can put endable loops into your program : one way is a little complicated to understand, but not very complicated to use.
Since loops are new to you, we will deal with the second method first.

8.2.1 The IF ..... THEN Statement

Earlier in this chapter we mentioned the two types of GOTO statement as being conditional and unconditional. You saw how the GOTO statement on its own can give continuous execution of part of a program. This is all an unconditional GOTO can do on its own. If you couple it to an IF .... THEN statement in the following generalized form:

    nn IF (variable) (condition)   THEN GOTO   mm

 - a new range of flexibility will be opened to you .

Let us now cover in detail the above line :

    1. nn and mm : both refer to line numbers that must be part of the program and different.

    2. Variable : the variable must have a value, but can be positive or negative.

    3. Conditions : this is the most important part of the line and needs the most care when writing the program. There are two parts to this part : firstly the relations (of which there are

four); secondly the variable or numeric expression or arithmetic expression. We'll clarify this with a four examples, using X as a variable to the left of the conditions :

```
IF X > Y THEN
IF X < 10 THEN
IF X < Y**2 THEN
IF X <= Y THEN
IF X >= Y THEN
```

These symbols (if you haven't seen them before) will be somewhat mystifying but you will understand shortly.

They are the relations, and are always read from left to right :

this symbol   >   means 'greater than'
this symbol   <   means 'less than'
and these   > = , < = mean 'greater than or equal to' and 'less than or equal to' respectively. If feel you may experience difficulty in remembering which is which, the following rules will help :

  1. Always read them from left to right.

  2. Consider the number of 'points' presented as you read it : e.g. < two here; >   one here.
   - then see that two is greater than one.

  4. THEN GO TO mm : only if the previous conditions are true will execution be transferred to the designated line number. If not execution 'falls through' to the next line.

In order to rewrite the program we previously entered we must start off a counter that will

count the number of times the loop has executed.
When the desired level has been reached, the
conditional GOTO will transfer execution outside
the loop.

It seems as if this is an extremely involved way
of achieving our desired results, but once you
see these in a program, it will become clearer.

Below is our new program for the one we showed
you earlier:

```
10 PRINT "ENTER YOUR NUMBER WHEN THE CURSOR
   CHANGES TO    L  "
20 LET B = 0
30 LET C = 0
40 INPUT A
50 LET B = B + A
60 PRINT A,B
70 LET C = C + 1
80 IF C <= 10 THEN GOTO 40
90 PRINT "PROGRAM TERMINATED"
```

Line 30 to set the counter to its initial value.

Line 70 increments the counter by 1 each time the
loop is executed.

There is a little backward thinking here that
sometimes can be quite useful when programming :

Line 80 does not transfer execution outside the
loop at all; it merely stops transferring
execution to the loop when the counter reaches
10.

Line 90 is not necessary, but can occassionally
be a nice touch.

8.2.2 FOR ...... NEXT Statement

As you may have noticed, the above method is not very compact, and although the verastibity of the program has been increased, its length has been increased also. Unfortunately it is not often possible to increase versatility without increasing length, but there are efficient and inefficient means of achieving the same ends. This is where FOR ...... NEXT looping comes in,

Firstly, the general forms of the statements :

    FOR (variable) = aa TO bb (STEP cc)
    .
    .
    .
and NEXT (variable)

- are very compact. The two lines above will replace lines 30, 70 and 80 in our previous program.

The FOR statement in more detail :

1. The (variable) must be unique to the loop in that it must not have its value changed at any point in the program. The (variable) in this case must be a single alphanumeric variable.

2. The lower case letter aa, bb, cc stand for integer values : aa is the initial value of the (variable); bb is the final value of the (variable); and cc is the value of the increments. If you wish to increment by 1, then it is not necessary to include STEP cc.

3. A FOR must always appear in a program with a NEXT , and it must be before the NEXT.

With the NEXT, the only rule is that it must

contain the same variable as the FOR statement.

Let us now put the FOR ...... NEXT loop to work, by rewriting the rewrite of the program we introduced at the start of the chapter :

```
10 PRINT  "ENTER YOUR NUMBER WHEN THE CURSOR
   CHANGES TO  L  "
20 LET B = 0
30 FOR N = 0 TO 9
40 INPUT A
50 LET B = B + A
60 PRINT A,B
70 NEXT N
80 PRINT "PROGRAM TERMINATED"
```

As you can see, this program is more compact than the previous one, and this is an advantage considering that memory space is limited.

Also , it is for less unwieldy than the previous method.

8.3   More IF ... THEN

There is more than one use for the IF ..... THEN statement; it can be used to transfer execution from one place in the program to another, depending upon the value of a variable being used. The following programs will serve to illustrate the uses of the IF .... THEN.

```
10 PRINT "DO YOU SEE ANY CLOUDS IN THE SKY?
   YES OR NO"
20 INPUT A$
30 PRINT "IS IT RAINING NOW? YES OR NO"
40 INPUT B$
50 IF NOT A$="YES" OR NOT B$="YES" THEN GOTO
   80
60 PRINT"IF YOU DO GO OUT TODAY, TAKE AN
   UMBRELLA..."
```

```
80 PRINT "IT'S A NICE DAY TO GO OUT ....."
90 PRINT "PROGRAM TERMINATED"
```

NOTE :
Line 50 introduces the NOT and OR logical operators, and they mean just as they say. If you have difficulty translating the statement, then consider it in pieces. The translation you should get is :

```
50 IF A$="NO" OR B$="NO" THEN GOTO 80
```

Follow the program through and make sure you can predict how it would run before emptying ZX81's memory and entering it.

```
10 PRINT "INPUT TODAY'S DAY?"
20 INPUT D
30 PRINT "INPUT THIS MONTH?"
40 INPUT M
50 IF D = 25 AND M=12 THEN GOTO 90
60 IF D = 1 AND M = 1 THEN GOTO 110
70 PRINT "TODAY'S DAY IS NEITHER CHRISTMAS DAY
       NOR NEW YEAR"
80 GOTO 120
90 PRINT"TODAY IS CHRISTMAS"
100 GOTO 120
110 PRINT"TODAY IS NEW YEAR"
120 STOP
```

This program shows the use of the IF .... THEN to sort out data and get the appropriate response from the computer. Lines 50 and 60 will transfer execution only if both expression on the left and right of the logical operator are true.

```
100 CLS
110 PRINT"INPUT THE ORIGINAL AMOUNT (IN $)"
120 INPUT A
130 CLS
```

```
140 PRINT"INPUT THE ANNUAL INTEREST RATE
    (PERCENTAGE)"
150 INPUT I
160 CLS
170 PRINT"INPUT NUMBER OF YEARS"
180 INPUT Y
190 CLS
200 PRINT"INPUT NUMBER OF TIMES A YEAR THAT
        INTEREST IS COMPOUNDED"
210 INPUT T
220 CLS
230 IF T=999 THEN GOTO 380
240 IF A(=0 OR I(=0 OR Y(=0 OR T(=0 THEN GOTO
    340
250 LET N=Y*T
260 LET R1=I/100/T
270 LET B=A*(1+R1)**N
280 PRINT"FINAL AMOUNT =$";B;" FOR ";Y;" YEARS"
290 PRINT
300 PRINT"DO YOU WISH TO CONTINUE? YES OR NO"
310 INPUT A$
320 IF A$="NO" THEN GOTO 380
330 GOTO 110
340 CLS
350 REM ERROR ROUTINE
360 PRINT"INCORRECT DATA. PLEASE RETYPE DATA?"
370 GOTO 100
380 STOP
```

This program will calculate the compound interest on an investment. It will also check for incorrect data, with line 240. This line is a series of logical operators, all of which must be true or an error message will be printed.

QUESTIONS

1. Rewrite the following program using IF...THEN instead of the FOR loop.

```
100 FOR I=1 TO 10
110 PRINT INT(RND*I)+1
120 NEXT I
```

2. Rewrite the following program using a FOR loop instead of the IF...THEN. It should be more compact.

```
100 LET S=0
110 PRINT "SCORE: ";S
120 LET S=S+1
130 IF S<10 THEN GOTO 110
```

3. Rewrite the following program without using any FOR statements.

```
100 FOR I=1 TO 5
110 FOR J=2 TO 6
120 PRINT I,J,I+J
130 NEXT J
140 NEXT I
```

CHAPTER 9 REM, STOP AND CONT

In this chapter, we will introduce you to a command and its accompanying statement, and another statement. These items are not closely related (in fact, one of them does precisely nothing in the running of a program), but have been put here so you can learnt of some fine detail, in computer programming.

From now on, when we give you an example of a program to enter, we will not tell you to use the command NEW to empty memory prior to entry. This is because it is good programming practice to empty memory prior to program entry.

9.1 Casual REMark

We will now tell you about the REMark statement. The function of the statement is to hold information or comments that may help other programmers viewing a listing of your program understand its operation. The important thing to remember is that the computer ignores these lines completely when excuting a program.

Consider the following generalised line :

   (line number) REM (Whatever you desire)

In the program below, you will see that of the seven lines it contains, only the final line will give output :

```
          10 REM THIS IS A SAMPLE PROGRAM THAT PROVES
             THE COMPUTER IGNORES.
```

```
20 REM LINES STARTING WITH REM STATEMENTS.
30 REM PRINT"THIS LINE WILL PROVE LINES 10
   AND 20 ARE FALSE"
40 REM LET A = 15
50 REM LET B = A/3
60 REM PRINT A,B
70 PRINT "PROGRAM TERMINATED"
```

Try running the program and observe the output :

   PROGRAM TERMINATED

As you can see, the only part of the program that produced any output or was operated on was the final line. This proves that the REM statement does as we earlier said.

9.2   STOPping your program
========================

You may not always wish to print "PROGRAM TERMINATED" at the end of a program, even though it can be a useful message. This is where the STOP statement comes in. Like many statement and commands in BASIC, its function is self explanatory. When the computer encounters this command, execution of the program will cease. You may use the statement as a program line or as equivalent of the BREAK command. The difference between STOP and BREAK is that STOP causes the computer to remember the next line due for execution, whereas BREAK causes execution to cease only. When using BREAK, a sign   D/(line number) will appear at the bottom of the screen. The line number is the next line for execution, and the 'D' is a flag to let you know the BREAK statement has been used. A report code '9' will be displayed instead of 'D' if the STOP statement is used as a program statement. Like

55

BREAK, STOP may be used at any time during execution.

The value of STOP is that the computer knows which line it was to execute next. Therefore, it is possible for program execution to be continued where it left off, and you may do this by using the CONT statement. After STOPping a program, type CONT to set the program running again.

In the case of very short programs, you will not have much opportunity to use BREAK and CONT from the keyboard, as the time taken to RUN the programs is very short. If you RUN a program that require you to enter a datum and you need time to consider your reply (or get a cup of coffee), then you may STOP the program.

Type in the following programs and use the CONTinue after the program is halted.

e.g (1) 10 PRINT "TESTING THE STOP STATEMENT"
         20 STOP
         30 PRINT " CONT STATEMENT DOES WHAT IT SAYS"

After the command CONT has used, your ZX81 will continue to excute the next statement. Firstly, your ZX81 will clear the screen and then display:-

   CONT STATEMENT DOES WHAT IT SAYS

e.g (2)   10 LET A=2
             20 LET B=4
             30 FOR I=A TO B
             40 INPUT C
             50 STOP

            60 NEXT I

:- After you enter a number, a message 9/50 will appear on the screen. If you use the statement CONT, your ZX81 will execute the NEXT statement. By adding 3 statements, you can check your input.
   Try it now !!!!

e.g (3)   10 LET A=2
               20 LET B=3
               30 LET C=A+B
               40 PRINT "THE SUM IS△";C
               50 STOP

:- If you use STOP at the very end of your program. The message with report code '9' will be displayed. But if you use then use CONT, your ZX81 will display   0/0 . This message is because there are no more program statements after the line 50.

QUESTIONS

1. What will be the report code when this program stops? And at which line will CONT restart the program?
100 PRINT "HELLO"
110 STOP
120 PRINT "HELLO THERE"

2. Where will this program stop?

100 PRINT "LINE 100"
110 REM STOP
120 PRINT "LINE 120"

CHAPTER 10 **ARRAYS**

As we said back in Chapter 1, one of the things computers can do very well is remember data. There is more than one way the computer can do this, one of which allows for easier manipulations than the other.

You may give each datum a variable name, but manipulation of individual data would be difficult as you would have to know what each variable name indicated, and this tends to defeat the purpose of putting them in the computer in the first place. In any case, the variable names would take up an excessive amount of memory.

The other way, is to tell the computer to reserve a piece of memory, divide this piece up into as many smaller pieces as required, give each small piece a small label, and finally name the large piece with one name. All this can be done with one line of program.

Phew!!

Now, the naming of this piece of memory takes only a small amount of space, and the labels are numeric, requiring no space at all as the computer counts its way along each little piece. This leaves you with far more room to hold data.

We call these large memory pieces ARRAYS.

When forming an array, there are two rules to remember in the naming :

   1. An array name must be a single letter.

2. No two arrays may have the same name.

As you can see, this gives you up to twenty-six possible arrays.

10.1 Forming Arrays : DIM statement

When the computer encounters the DIM statement — this need not be at the start of the program, provided it is before the array is to be used — it will perform the feats we told you about earlier.

Below is an example of a DIM statement that will reserve twelve spaces for data, under the general name M. These spaces will be numbered from 1 to 12, with initial values of zero.

        DIM M(12)

Put into a diagramtic form array M would look like :

M

1	2	3	4	5	6	7	8	9	10	11	12

If you wish to call or perform an operation on the contents of an individual part — or element — of an array then you simply state the array name with the element in brackets afterwards. Alternatively if you have variable that has an integer value, then this may be used in place of the element number.

Below is a program that allows you to load data into array M, and as a check, it will then print out the data you have loaded.

```
 1 DIM M(12)
10 REM : THIS PROGRAM IS A TRIAL TO
ILLUSTRATE THE USE OF ARRAYS.
20 PRINT "ENTER YOUR DATA"
30 FOR I=1 TO 12
40 INPUT M(I)
50 NEXT I
60 REM : NOW DISPLAY DATA
70 FOR I= 12 TO 1 STEP -1
80 PRINT "DATAM ";I;"=";M(I)
90 NEXT I
```

NOTE the program is controlled by two FOR.......NEXT loops.

Below is a listing of a program which is called a bubble sort. The aim is to allow you to input a number of values into an array, and then to sort them into in increasing order. Examine the program carefully and make sure you understand it before typing it into your ZX81. Then run the program to see that it works.

```
10 DIM S(10)
20 PRINT " HOW MANY NUMBERS ?"
30 INPUT N
40 IF N < 10 THEN GOTO 50
42 PRINT " THE MAXIMUM NO. <= 10"
45 GOTO 20
48 CLS
50 PRINT " INPUT YOOUR NUMBERS ?"
60 FOR I=1 TO N
70 INPUT S(I)
80 NEXT I
90 FOR I=1 TO N-1
100 FOR J=1 TO N-I
```

```
110 LET X=S(J)
120 LET Y=S(J+1)
130 IF X <= Y THEN GOTO 160
140 LET S(J)=Y
150 LET S(J+1)=X
160 NEXT J
170 NEXT I
180 CLS
190 PRINT "THE NUMBERS ARE IN INCREASING ORDER"
200 FOR I=1 TO N
210 PRINT S(I)
220 NEXT I
230 STOP
```

Here is another example of a program using arrays. It illustrates the way the elements can be manipulated.

This is a puzzle to test your powers of logic! You start with 9 numbers in an array, in random order. The aim is to get them into ascending order in the least possible number of moves. The only way you have to rearrange them is to reverse the order of some elements.
For instance, if you have
    1 5 4 9 2 3 8 7 6
and reverse 3 of them you will have
    4 5 1 9 2 3 8 7 6
Then reverse 6 of them, and you will have
    3 2 9 1 5 4 8 7 6

NUMBERS

```
10   LET B=9
20   DIM A(B)
30   FOR C=1 TO B
40   LET A(C)=C
50   NEXT C
60   FOR C=B TO 2 STEP -1
```

```
 70 LET D=INT(C*RND)+1
 80 LET E=A(C)
 90 LET A(C)=A(D)
100 LET A(D)=E
110 NEXT C
120 LET E=0
130 GOSUB 320
140 PRINT "REVERSE?"
150 INPUT F
160 IF F=0 THEN STOP
170 IF F<=B THEN GOTO 200
180 PRINT B;" MAXIMUM"
190 GOTO 140
200 LET E=E+1
210 FOR D=1 TO INT(F/2)
220 LET G=A(D)
230 LET A(D)=A(F-D+1)
240 LET A(F-D+1)=G
250 NEXT D
260 GOSUB 320
270 FOR D=1 TO B
280 IF A(D)<>D THEN GOTO 140
290 NEXT D
300 PRINT E;" MOVES"
310 STOP
320 CLS
330 FOR D=1 TO B
340 PRINT A(D);" ";
350 NEXT D
360 PRINT
370 RETURN
```

Lines 30 to 50 initialize the array variables to the numbers 1 to 9. Using the FOR loop is a very compact and neat way of handling arrays, and it is used a lot.

Lines 60 to 110 rearrange the elements of the array in a random order. Note that lines 80 to 100 swap the values of elements C and D in the array.

Lines 210 to 250 reverse the elements as directed

by the operator.
The subroutine in lines 320 to 370 PRINTS the array.

10.2 Two Dimensional Arrays

Section 10.1 was concerned only with generating arrays that consisted of lists of data or one dimensional arrays. As the one dimensional array is related to a straight line, so the two dimensional array is related to a flat plane as in a piece of paper.

Consider the following diagram :

```
        1   2   3   4   5
      +---+---+---+---+---+
    1 |   |   |   |   |   |
      +---+---+---+---+---+
    2 |   |   |   |   |   |
      +---+---+---+---+---+
    3 |   |   |   |   |   |
      +---+---+---+---+---+
    4 |   |   |   |   |   |
      +---+---+---+---+---+
```

It shows a grid representation of a two dimensional array. The numbers down are termed 'lines' and the numbers across are termed 'columns'.

The use of two dimensional arrays is that a single value may be held by a combination of two variables. Should
you wish to write a program for a 'grid' game, then use of a two dimensional array will make program writing easier.

You can generate a two dimensional array in a similar fashion for one dimensional arrays, like so :

        10 DIM A(4,5)

- where the first number denotes the number of lines and the second denotes the number of columns.

Loading values into the array is little different from before, you may use a FOR....NEXT loop like so :

         1 DIM A(4,5)
        10 FOR N=1 TO 5
        20 FOR M=1 TO 4
        30 INPUT A(M,N)
        40 NEXT M
        50 NEXT N
        60 STOP

As you see, there are two FOR....NEXT loops, one within the other. The loop with lines 20 to 40 is 'nested' within the loop with lines 10 to 50. You can nest as many loops within each other as you like, provided they are completely contained within each other. If, for instance the program lines 40 to 50 were reversed then it would be considered illegal by the computer.

To change just one value in the array, use a LET statement, with the co-ordinates of desired element then its value, like so :

        10 LET A(2,3)=8

10.3 Three and more dimensions

You may have arrays with as many dimensions as you wish although three is usually the maximum used.

In the same manner as we related one and two dimensional arrays to a straight line and a flat plane, so we can relate a three dimensional array to a box. The box will be divided into an equal number of parts, the same number as there are elements in the arrays.

Generating such an array is simular to the generation of a two dimensional array, as is loading.

It is possible to visualize a four dimensional array as a volume of space over a period of time, but, it is very difficult to visualize arrays with dimensions beyond this.

10.4   String Variable Arrays.

Just as you have arrays for numeric variables, so you can have arrays for string variables. There are some differences as you see from the general form of generating the array :

   DIM (String Array Name)$(number of elements,length of elements)

1. String Array Name must consist of a single letter followed by the $ sign.

2. In a string array, all the strings have the sames fixed length. Therefore, the DIMension line has a extra number(the last one), to specify this length.

3. A string array name must be exclusive, and a program cannot contain a string variable with the same name, i.e.

           A$(6,5)   or   A$

but not both.
This differs from numeric arrays, where the names
do not have to be exclusive.

4. You can use 'slicing' techniques – discussed
in the previous chapter – may be used in string
arrays. For example :

```
10 DIM A$(5,10)
20 LET A$(1)="12345678"
30 PRINT A$(1),A$(1,7)
40 PRINT A$(1),A$(1,2 TO 5)
 50 STOP
```

Display :   12345678   7
             12345678   2345

10.5   Array Storage

It is possible to store the data contained in
array, but not directly. To store data, you must
go through the following steps :

1. Set up an array and load in your values using
a program such as you were shown in section 10.1
and 10.2.

2. Edit out this program. Do not use NEW, this
will also clear the array from memory.

3. Type in your program that uses the data
stored in the array, and then SAVE it (we will
show you how to do this later). The effect of
this will be to store the program and the
variables contained in the array.

4. When LOADing the program back, the array and
its variable will also be LOADed.

5. When you execute the program, do not use RUN, or the variables will be cleared. Use the statement GOTO nn instead. Where nn is the first line number of your program.

QUESTIONS

1. Write a program that sets up a one-dimensional numeric array of 10 elements. Store the numbers 1 to 10 in the array:- 10 in the first element, 9 in the second, etc. Use a FOR loop.

2. Write a program that sets up a two-dimensional numeric array with 5 by 5 elements. Store the numbers 0 to 4 in it as follows:-
```
    0 1 2 3 4
    1 0 1 2 3
    2 1 0 1 2
    3 2 1 0 1
    4 3 2 1 0
```
Hint:- The number in X(I,J) is ABS(I-J)

3. Set up a string array that contains the days of the week, so that D$(1)="SUNDAY". Using this array, write a program which will INPUT a number betwwen 1 and 7, then PRINT the corresponding day.

CHAPTER 11 **SUBROUTINES**

You are now at the stage where you have learnt enough of the ZX81's vocabulary to write some quite useful programs. In this chapter, we will introduce you to subroutines, so your program can be more efficiently written and structured.

A subroutine is a part of a program that does a task a number of times during the running of a program, but at different stages. For example, if you wrote a program that required ten random numbers at four different times, then it would be laborious to type in the same series of lines four times. An easier way would be to type in the lines once and transfer execution to those lines whenever the numbers were needed.

This is the function of GOSUB, the statement that transfers execution to the subroutine. The statement appears generally as :

      GOSUB nnn

- where nnn is the first line of the subroutine.

GOSUB always have the RETURN statement at the end of the subroutine. The effect of RETURN is to return execution to the program line that comes immediately after the GOSUB statement.

You may have noticed the similarity between GOSUB and an unconditional GOTO. They are similar initially, but the transferring of execution back is what makes the GOSUB superior to GOTO in this case. When the ZX81 encounters a GOTO, it will not remember which line comes next in the program and cannot transfer back unless at the end of the subroutine there is another unconditional GOTO.

The following program will illustrate how the GOSUB and RETURN statements work :

```
 10 REM : THIS WILL ILLUSTRATE GOSUB STATEMENT
 20 PRINT "THIS IS THE MAIN PROGRAM"
 30 PRINT "I AM ENTERING A SUBROUTINE NOW"
 40 GOSUB 100
 50 CLS
 60 PRINT "I HAVE RETURNED TO THE MAIN PROGRAM"

 70 STOP
100 REM : THE START OF SUBROUTINE
110 PRINT
120 PRINT "THIS IS THE START OF SUBROUTINE "
130 PRINT "CHECK THAT WHEN THIS SUBROUTINE HAS FINISHED."
140 PRINT " I WILL BE RETURNING AND EXECUTE LINE 50"
150 PRINT "AND THE SCREEN WILL TURNS BLANK........"
160 RETURN
```

The diagram below is another way of showing how the two statements work :

```
 10 REM.....
  .
  .
  .
 50 GOSUB 1000
  :
  :
  :
 90 GOSUB 1000
100 STOP
```

```
1000 REM SUBROUTINE 1000
       :
       :
       :
1150 RETURN
```

The program below will generate random integer from 0 to 1 to simulate the tossing of a coin. It contains two subroutines to show how they can be used. Note that for efficiency, only one subroutine need be used, as the two in the program can easily be combined. Of course, a subroutine is not strictly needed, as the program is very short; they have only been used for demonstration purposes.

```
 10 REM : THIS PROGRAM WILL PRINT OUT THE
         RESULTS OF TEN SIMULATED COIN TOSSES.
 20 RAND 0
 30 FOR I= 1 TO 10
 40 GOSUB 100
 50 GOSUB 200
 60 NEXT I
 70 STOP
100 LET A=INT(RND+0.5)
120 RETURN
200 IF A=1 THEN GOTO 230
210 PRINT "TOSS ";I;"= TAIL"
220 RETURN
230 PRINT "TOSS ";I;"= HEAD"
240 RETURN
```

By writing a program as a series of subroutines, you make it more easily read :

GOSUB 100 signifies the place in the program where the
    simulated tossings are taking place.

GOSUB 200 signifies where the output will be performed.

The above program can be rewritten to make it more efficient (as we said earlier), by deleting line 50 and inserting :
    110 GOSUB 200

Try it, you should see no difference in the output.

QUESTIONS

1. Write a subroutine that prints the numbers in an array of 10 elements. Use this subroutine in the bubble sort program in chapter 10 to print the array every time two numbers are exchanged, as well as at the end of the program. Then you will be able to watch the numbers being sorted.

2. What happens if you don't RETURN from subroutines but use a GOTO statement instead? Try the following program:-

100 GOSUB 1000
1000 SCROLL
1010 PRINT "1234567890"
1020 GOTO 100

To understand what is happening, remember that a GOSUB statement puts the return address on the stack, and the RETURN takes the top address off the stack.

71

CHAPTER 12 **CHARACTERS**

'Characters' is a word used to encompass letters, digits, punctuation marks and things that can appear in strings : and they are all on the ZX81's keyboard.

There are 256 characters in your ZX81, and they are mostly single symbols such as A, B, C, X, etc. Some characters represent whole words (LET, STOP, **, etc.) and are called tokens. All the characters and tokens have their own number in the range 0 to 255, called a 'code'. By using the functions CODE and CHR$ it is possible to convert a character to its code and vica versa.

We will now cover these functions in more detail.

1. CODE : is applied to a string. This function will give the code of the first character in the string, or zero if the string is empty. As an example of the use of this function, consider the program below, and then run it :

```
10 DIM A$(10,10)
20 PRINT "INPUT A STRING OF LENGTH <=
   10"
30 FOR J=1 TO 10
40 INPUT A$(I)
50 NEXT I
60 CLS
70 PRINT "STRING","CODE"
80 FOR J=1 TO 10
90 PRINT A$(J), CODE A$(J)
100 NEXT J
110 STOP
```

Line 10 will define a piece of memory sufficient

hold ten strings of length ten characters.

Line 80 to 100 will print out the strings you have input, together with the code of the first character.

2. CHR$: is applied to a number. This function give a single character string whose code is that number. The program below will print out each of the characters of the ZX81 in sequence. This will happen extremely quickly, but do not worry, later on we will show you how to slow it down.

```
10 LET A = 0
20 PRINT CHR$ A;
30 LET A = A + 1
40 IF A < 256 THEN GOTO 20
```

Appendix A contains a table which lists the characters together with their codes.

In the running of the program, you may have seen the characters written in black on white or inverse video. These characters may be accessed from the keyboard as follows :

1. Press GRAPHICS — see the cursor G appear. This signifies the computer is in graphics mode.

2. Press the desired symbol. It will be displayed in inverse video.

3. To turn the inverse video off, press either GRAPHICS or NEWLINE.

When you ran the previous program, you will have noticed certain patterns were displayed. These are included to allow you to be more creative with your output. You may also access these from the keyboard like so :

1. Press GRAPHICS to put the computer in graphics mode.

2. Press SHIFT and the desired graphic symbol.

3. To resume normal mode, press NEWLINE or GRAPHICS.

Look carefully at the keyboard, and you will see that all the graphics symbols are located on the keys that are also 'token' keys. As tokens have no inverse, you get a graphics symbol when using the above steps.

Try the following program, which is designed to be used in horizontal bar charts - using grey and black.

```
100 PRINT "INPUT A"
110 INPUT A
120 PRINT "INPUT B"
130 INPUT B
140 CLS
150 IF A<30 AND B<30 THEN GOTO 180
160 PRINT "A AND B MUST BE LESS THAN 30"
170 GOTO 100
180 FOR I=1 TO A
190 PRINT "■"; (graphic A)
200 NEXT I
210 PRINT TAB 31;"A"
220 FOR I=1 TO B
230 PRINT "□" (graphic space )
240 NEXT I
250 PRINT TAB 31;"B"
```

Observe line 210. This line contains a TAB function. The purpose of this function is move the PRINT across the screen to the specified column. When using TAB, you normally wish to suppress the carraige return (or NEWLINE), and this can be done by using a semi-colon (;). Thus the general form in which TAB appears is :

    PRINT TAB n;

You may TAB as far across the screen you like, but if you use several TAB function to the same line, then you must use increasing numbers for each TAB. Using decreasing numbers will cause the ZX81 to become confused, and it will respond by moving on to a newline.

Note : Your ZX81 has only 32 columns across, so if you TAB more than 32 in total, then it will move to the next line.

i.e.   PRINT TAB 33; will see the output commence one space in, one line down.

Here is another program that will show what you can do with graphic characters. It is a game program called TELEPORT.

Captain Kirk is lost! You know that he is on one of the displayed stars, but which one? If you don't locate him, and teleport him back to the ship within 5 star-days, he will starve.

When prompted by the computer, input the X-coordinate and then the Y-coordinate of your guess. The computer will tell you if your guess (coordinates A,B) is less than, equal to, or greater than Captain Kirk's position (coordinates X,Y).

TELEPORT
(c) by Clifford Ramshaw

```
100 LET X=INT (RND*8)
110 LET Y=INT (RND*8)
120 FOR I=PI/PI TO 3+RND*6
130 PRINT AT RND*7,RND*7;"*"
140 NEXT I
150 PRINT AT Y,X;"*";AT 3,9;"▆▆▆"
                (3 * graphic F)
160 FOR T=PI/PI TO 5
170 PRINT AT 4,9;"X,Y? TIME:△";T
180 INPUT A
190 INPUT B
200 PRINT,"X ";
210 PRINT (">" AND A<X)+("<" AND A>X)+("=" AND
    A=X);" A"
220 PRINT,"Y ";
230 PRINT (">" AND B<Y)+("<" AND B>Y)+("=" AND
    B=Y);" B"
240 IF A=X AND B=Y THEN GOTO 270
250 NEXT T
260 PRINT "TOO LATE CAPT. KIRK IS DEAD";CHR$ 300
270 PRINT AT Y,X;"■" (graphic A);AT PI-PI,10;"■"
    (graphic A),TAB 9;"▆▆▆" (graphic T, space,
    graphic Y)
280 PAUSE 100
290 PRINT AT Y,X;"*"; AT PI-PI,10;"O",TAB 9;"▆Z▆"
    (graphic T,inverse Z,graphic Y),TAB 9;"▆▆▆"
    (graphic T, space,graphic Y); AT 4,8;"BEAMED
    UP IN ";T;" TRIES"
```

Lines 270 to 290 simulate teleporting Captain Kirk back to safety, using the graphic characters you have just met.

If you are having trouble understanding line 210, an equivalent way of writing that would be

```
210 IF A<X THEN PRINT ">";
212 IF A>X THEN PRINT "<";
214 IF A=X THEN PRINT "=";
216 PRINT " A"
```

Similarly for line 230.

QUESTIONS

1. Write a program which asks if you want to stop. If the first letter of the answer is "Y" then it stops; otherwise it goes back to the beginning. There is more than one way of writing this.
This is a useful program segment for interactive programs that have the option of repeating.

2. Rewrite the bar chart program to draw vertical bars.

CHAPTER 13 TWO SPEED COMPUTER

Up until now, you have been using your ZX81 at its slowest speed,. Even so, it has been obeying its instructions at quite a rapid pace. At this 'slow' speed, the computer is able to display information and compute simultaneously.

Selection of speed is by the statements FAST and SLOW for fast and slow speeds respectively. The difference between SLOW and FAST is a factor of about 4 times.

Sometimes, you may have a program that requires considerable computation with little output, and this is where the faster speed becomes useful.

Also, when typing in a long program : you may have already noticed how long it takes for each statement to be displayed on the screen.

Try the following program, but first type FAST then NEWLINE, to put the computer in fast mode.

```
10 FOR N = 0 TO 255
20 PRINT CHR$ N;
30 NEXT N
```

NOTE that the program did not display anything until the end of the program.

Now try the next program below. Observe the computer displays while it is in INPUT mode, as it is waiting for data.

```
10 INPUT A
20 PRINT A
30 GOTO 10
```

When you wish your computer to be put back to
SLOW so you can get some output, use the same
method as for FAST, but use the statement SLOW
unstead of FAST.

As both FAST and SLOW are statements, they can be
used as part of a program. Below is an example
containing the two statements :

```
        10 SLOW
        20 FOR N = 1 TO 64
        30 PRINT "THIS PROGRAM ILLUSTRATES THE
           DIFFERENCE BETWEEN FAST AND SLOW"
        40 IF N = 32 THEN FAST
        50 NEXT N
        60 GOTO 10
```

N.B.   You may also set the computer to SLOW
(from FAST) by turning it off, then on. This has
the added advantage of clearing the computer's
memory at the same time.

CHAPTER 14 **OUTPUT**

This chapter will cover the ways in which you may vary the output (or display data) from your ZX81. Some items in this chapter you will already be familiar with, but for completemess we will include them here also.

14.1   PRINTing Output :

The PRINT statement together with other characters and functions may give varied output in five ways :

1.   A line containing PRINT only will simply cause a blank line to be output. The computer interprets this to mean 'print nothing', and this is what happens.

2.   A semicolon (;) in a line containing a PRINT statement will supress the carriage return (NEWLINE). The carriage return is the movement of the cursor to the left hand column on the next line. In effect, you get all output on the same line (provided there is room) without spaces.

3.   Commas after the PRINT will result in the output being divided into two columns. With more than two variables or items to be output, your ZX81 will more than one line.

4.   When you use PRINT AT, you are effectively telling your ZX81 where to PRINT.
Consider the example below :

       PRINT AT 11, 16, "*"

             Line   Column

In the above example, your ZX81 will place an asterisk on line 11, column 16. This is the middle of the screen. Lines are numbered from 0 (top) to 21 (bottom). Columns from 0 (extreme left) to 31 (extreme right).

5. TAB was introduced to you in the previous chapter, and is used to move the PRINT position to the desired column number.

In conjuction with the five facilities above, there are four points to remember :

1. For the items AT and TAB, it is best to terminate with semicolons - in this case, you will remember where the last PRINT position is.

2. Both AT and TAB , are accessed by using the function key.

3. You cannot make your ZX81 print on the bottom two lines, as they are reserved for commands, INPUT, and so on.

4. When using AT, you may PRINT anywhere (except the bottom two lines), even where there is already some output; that already there will be overwritten.

This program illustrates how you can format the way your output appears on the screen. It PRINTs numbers in a list with the decimal points aligned. Don't input numbers with more than 8 digits before the decimal point, as the program will not align the decimal points then. If there are more than 2 decimal places, the number will be truncated.

```
100 INPUT M
110 LET N=INT(M*100)/100
120 LET X=LEN STR$(N*100)
```

```
130 IF ABS N<1 THEN LET X=X+1
140 PRINT TAB 20;
150 GOSUB 1000
160 PRINT N;
170 IF N-INT N=0 THEN PRINT ".0";
180 IF N*10-INT(N*10)=0 THEN PRINT "0";
190 PRINT
200 GOTO 100
1000 FOR Y=1 TO 10-X
1010 PRINT " "
1020 NEXT Y
1030 RETURN
```

14.2   Screen Manipulation

There are two statements available for manipulating the screen. They are SCROLL and CLS (CLear Screen).

Firstly, CLS will clear the screen completely and then put the print cursor to the zero line, zero column position (top left- hand corner of screen). The program below will illustrate the use of CLS :

```
10 FOR N = 1 TO 22
20 PRINT " TEST CLS STATEMENT : LINE ";N
30 NEXT N
40 CLS
50 PRINT "AS YOU SAW THE SCREEN WAS CLEARED"
```

The function of the SCROLL statement is to move the whole display from line number 0 to 20 up one line. Line 0 is removed from the display, and the next line is placed on the bottom of the current display. You need not have the whole screen filled as the example below shows :

```
10 CLS
20 FOR N = 1 TO 25
30 PRINT "SCROLL DEMONSTRATION, LINE";N
```

```
40 IF N < 5 THEN GOTO 60
50 SCROLL
60 NEXT N
```

14.3 PLOTting and UNPLOTting

In this section, we will introduce you to the PLOT and UNPLOT statements. With these statements, it is possible to draw graphs, pictures and in fact, anything you wish. Clearly, these statements are very powerful. Before we show you the use of these statements, it will make things more clear if we show you how your ZX81 generates output on the screen.

Firstly, the screen is divided into pieces just big enough for any character it positions. Since the screen can take 32 characters across, and 22 down, there are 32 x 22 possible positions for characters, or 704 positions.

Athough each character is made up from a matrix of 8 x 8 dots = 64 dots total, for PLOT and UNPLOT, the spaces normally taken up by characters are divided up into a 2 x 2 matrix, each element of the matrix is called a pixel. Each pixel, therefore consists of 16 dots. By using the two statements, each pixel may be turned off or on as desired.

You may choose your pixel by determining its x co-ordinate and y co-ordinate. The x co-ordinate is how far from the extreme left hand column the pixel is; where the y co-ordinate is how far from the bottom of the screen the pixel is. The co-ordinates are separated by a comma, just as in cartesian co-ordinates. Below is a representation of the screen with the co-ordinates for each corner included :

```
(0,43)                    (63,43)




(0,0)                     (63,0)
```

To turn on a pixel (= black) use the PLOT statement and follow it with the co-ordinates of desired pixed (in brackets, of course). Turning off a pixel is done in much the same way, but the UNPLOT statement is used.

Examine the following program. The PAUSE statement is new, but is simply a way of making the computer slow down the rate at which the computer executes a program. When the PAUSE statement is encountered, the computer pauses in execution for a while. This statement will be covered in more detail in the next chapter. Another new statement is POKE . The function of this statement is to enter a single byte of information - a character code (range 0 to 255) - in the address (or memory position) indicated by the first number of the statement.

It sounds confused and not very much use, but the statement will be fully explained in another chapter.

```
10 LET X = INT(RND * 64)
20 LET Y = INT(RND * 64)
30 PLOT X,Y
40 PAUSE 50
50 POKE 16437,255
60 UNPLOT X,Y
80 GOTO 10
```

This program plots a print randomly each time you press NEWLINE , and unplots it after about one second. The delay is due to the line 40 and 50.

PLOT can also be used to plot graphs, as the following program demonstrates.

```
100 FOR Y=0 TO 60
110 PLOT Y,(SQR Y)*5
120 NEXT Y
```

The second coordinate is multiplied by 5 merely to fill the screen better. Next you can add axes, and you will have a graph.

```
100 PRINT AT 21,21;"40"
110 PRINT AT 21,11;"20"
120 PRINT AT 21,1;"0"
130 PRINT AT 15,0;"2"
140 PRINT AT 10,0;"4"
150 PRINT AT 5,0;"6"
160 FOR Y=0 TO 60
170 PLOT Y+2,(SQR Y)*5+2
180 NEXT Y
```

QUESTIONS

1. Write a program which plots the graph of the LN function.
You could also write programs which plot any of the other functions.

2. Write a program which draws a train which moves across the screen. You will need to consider what the train will look like, and also how you are going to blank out the old train before printing the new one.

85

CHAPTER 15 INKEY$ AND PAUSE

In this chapter you will be introduced to two new and useful statements, PAUSE and INKEY$.

15.1 PAUSE

As you saw in chapter 14, the PAUSE function can be used to slow down the execution of a program. If you remember back to chapter 12, we showed you how to print out all the characters of your ZX81. We also told you that later on, we would show you how to do it more slowly. This is one of the functions of the PAUSE statement.

The general form of PAUSE is :

    PAUSE n

- where n is the amount of time you wish to pause, in terms of how many frames your television shows in each second. Most televisions show fifty frames/second, so PAUSE 50, will halt execution for one second. The upper limit of n is 32767, which corresponds to just under eleven minutes. If you use a value greater than 32767, then the computer will PAUSE forever.

When the computer is PAUSEing, if you press any key other than SPACE or pound sign, the pause will be cut short. Pressing SPACE or pound sign will result in a break in the program as if BREAK had been typed. After the computer has finished PAUSEing the screen will flash or blink.

NOTE FOR USERS WITH 'OLD' ROMS: If you are running your ZX81 in FAST mode, or using a ZX80 with the old 8K ROM, then a line containing the

PAUSE, may need to be followed by :

    POKE 16437,255

- or your program may be erased, even though the PAUSE appears to be operation normally. This does not apply to the new 8K ROM - see note at beginning of book.

With the PAUSE, it is possible to program your ZX81 as a clock, although the accuracy will not be absolute. To obtain better value for PAUSE. The loss of accuracy is due to the time taken for the ZX81 to perform the other parts of the program.

However, you could try the following program :

```
10 REM : DRAW THE CLOCK
20 FOR I = 1 TO 10
30 PRINT AT 10-10*COS (N/6* PI ),10+10*SIN
   (N/6*PI );N
40 NEXT I
45 REM : DISPLAY THE TIME BY PLOTTING A
   SINGLE DOT ON THE EDGE WHEN 1 SECOND IS UP
50 FOR T=0 TO 10000
60 LET A = T/30 * PI
70 LET SX = 21 + 18 * SIN A
80 LET SY = 22 + 18 * COS A
90 PLOT SX,SY
100 PAUSE 42
110 UNPLOT SX,SY
120 NEXT T
```

15.2 INKEY$

One of the very useful functions of your ZX81 is INKEY$. This function scans the keyboard to see if any keys are being pressed. If a key is being pressed, then the result is the character of that

key, otherwise the result is an empty string.
Also note that the control characters do not have
their usual effect.

Below is a program that uses INKEY$, to make the
computer behave like a typewriter.

```
10 IF INKEY$ <> "" THEN GOTO 10
20 IF INKEY$ = "" THEN GOTO 20
30 PRINT INKEY$;
40 GOTO 10
```

Line 10 waits for you to release a key.
Line 20 waits for you to press a key.

You must remember that INPUT and INKEY$ are not
the same : since INKEY$ does not wait for you,
there is no need to press NEWLINE

The program below may give you some fun, but try
to find out what it does before typing it in and
running :

```
10 IF INKEY$ = "" THEN GOTO 10
20 PRINT AT 11,14;"OUCH"
30 IF INKEY$ <> "" THEN GOTO 30
40 PRINT AT 11,14;"▲▲▲▲"
50 GOTO 10
```

INKEY$ is very useful in game programs with a
graphics display. The following game is an
example.
You are in control of a spaceship engaged in
combat with an enemy spacehip. You must try to
destroy as many enemy spaceships as you can. If
you get destroyed, you will get another chance to
prove your skill as a fighter. After you have
lost three ships, you will be sacked - spaceships

cost money!!
You use the keys 6 and 7 to move up and down (in the direction of the arrows). The key 1 fires at the enemy. All of these are checked using the INKEY$ function.

```
100 LET J=3
110 LET S=-1
120 LET D=INT(RND*28)+3
130 LET S=S+1
140 PAUSE 20
150 LET H=11
160 LET L=INT(RND*20)
170 CLS
180 PRINT AT H,0;" O" (inverse =,O)
190 PRINT AT L,D;"O " (O,inverse =)
200 IF INKEY$="1" THEN GOTO 250
210 IF RND<.4 AND ABS(H-L)<2 THEN GOTO 300
220 LET H=H-(H>0)*(INKEY$ ="7")+(H<20)*(INKEY$
    ="6")
230 LET L=L+INT(RND*3)-1+(L<0)-(L>20)
240 GOTO 170
250 FOR A=2 TO D
260 PRINT AT H,A;"-"
270 NEXT A
280 IF H=L THEN GOTO 120
290 GOTO 210
300 FOR A=1 TO D
310 PRINT AT L,D-A;"-"
320 NEXT A
330 IF H<>L THEN GOTO 220
340 LET J=J-1
350 PRINT "YOU ARE HIT" (inverse)
360 IF J THEN GOTO 120
370 PRINT "GAME OVER" (inverse)
380 PRINT "SCORE ";S
390 PAUSE 50
400 RUN
```

QUESTIONS

1. In FAST mode, the computer displays the screen during a pause. Use this fact to display a random character for a second, then wait for the user to input which letter it was. Decrease the length of the pause as the user's score gets higher.

2. Write a program which draws a picture of a man. Use the keys with the arrows to move the man arround the screen, checking if the keys are being pressed with INKEY$. Remember to check that the man is not moving off the screen.

CHAPTER 16 SAVING PROGRAMS

16.1 The SAVE statement

By now you will have realized that if you turn off your ZX81, you lose the program and all the variables that were in the computer at the time. Typing in a program every time you want to use it is a slow and tedious task, so the ZX81 enables you to save the program on cassette. Then, next time you want that program, you only have to load it back from the cassette.

To save programs in this way, you will need a cassette recorder, and some cassettes. The recorder must have a microphone and an earphone socket, preferable 3.5mm jack sockets. You will also need a lead to connect the computer and the recorder. Another useful feature that your recorder may have is a tape counter, but this is not really necessary.

Type in a short program to practice using the SAVE command. You will need to practice before you save any long programs, otherwise you may lose the program. A one line program will do to start with.

Find a part of the cassette that is blank. Connect the microphone socket of the recorder to the socket marked 'MIC' on the computer.

Now you are ready to save the program. You must give the program a name. The name may have up to 127 characters, but you should not use inverse

characters, because the computer attatches a
special meaning to inverse characters i program
names. A possible name would be "PROGRAM"

To save this on tape, type in
    SAVE "PROGRAM"
but don't press NEWLINE yet.

Start the cassette recorder recording. Remember
to push down the record button as well as the
play button.

Press NEWLINE

You will now see a grey pattern on the television
screen. This is a 'lead-in' before the computer
sends the actual program to the recorder. After
the grey pattern, which lasts for about 5
seconds, you will see a pattern of black and
white stripes. The computer is now sending your
program to the recorder. When this has finished,
the computer should report back with 0/0. You can
stop the recorder now.

Wind back the cassette. Play the program that you
have just recorded. You will hear a soft buzz
(the lead-in) followed by a loud, high pitched
buzz (the program). If this is not what you hear,
you have done something wrong. Check the
connection between the computer and the recorder.
Or could you possibly have forgotten the record
button?

It is not just the program that is saved, but all
the variables as well. If you type a command

    LET A$="STRING"

before you save the program, then after you have

loaded the program, A$ will be defined, although there is no program statement defining it.

16.2 Loading the program

Firstly, position the cassette at the beginning of the program you have recorded (in the lead-in). This is where a tape counter can help if your recorder has one.

Connect the EAR socket on the computer to the earphone socket on the cassette recorder.

Turn the volume to approximately three quarters volume. If your cassette recorder has tone controls turn the treble high and the bass low.

Type in
    LOAD "PROGRAM"
without pressing NEWLINE yet.

Start the tape recorder playing, and press NEWLINE.

You will see black and white patterns on the television screen, though a bit different to the patterns you saw when recording. When this has finished, the computer should stop with report 0/0.

If this does not happen, most likely you have the volume wrong. If the lead-in is noisy, then turn down the volume a bit, otherwise turn it up. If the lead-in is noisy, then the computer will not realize that it is meant to be silent.

There are other things that can make the lead-in noisy.

1. The heads on your cassette recorder may be

dirty. Clean them, and try again. (Methylated spirits on a cotton bud works well if you haven't head cleaning fluid.)

2. Use computer quality cassettes - C10 or C12 for instance.

3. Old recorders are often noisy. Using a better quality cassette may help.

4. Don't have both sockets connected at once. Use only the microphone socket while saving, and only the earphone socket while loading.

5. Move to a wooden table if you are using a table with metal parts. Lifting the computer off the table does not seem to work as well.

6. Try running the recorder on batteries to eliminate mains hum.

7. Expensive, stereo recorders do not seem to work as well as cheap mono recorders.

You may have a program on tape, but you can't remember what you called it. You can still load it, although the computer is more sensitive to volume if you don't know the name, by using the command
      LOAD ""
and proceeding as before.

When you give a name, the computer looks throught the tape until it finds a program with that name. If there is not one there, you can stop it looking with the BREAK key. If there is no name, the computer will load the next program on the tape.

16.3 Using SAVE in programs.

The effect of using SAVE as a program statement is to automatically start the program running from the line after the SAVE statement after it is loaded.

For example, type in this program.

```
100 SAVE "PROGRAM"
110 FOR I=1 TO 20
120 PRINT I
130 NEXT I
```

Save the program, then load it back again. It will begin to run as soon as it has finished loading.

If you list the program, you will find that the M in line 100 has changed to inverse. The computer uses this as a marker. That is why you should not use inverse characters in the program name.

CHAPTER 17 TOP DOWN PROGRAMMING

In your programming, you will already have had problems translating your flowchart into BASIC. When you come to a decision box, which branch do you write down first? The larger your program, and the more decisions that have to be made, the harder this problem is to solve.

The problem arises because a flowchart is a two dimensional representation of the program. The arrows (which show possible paths of execution) go up and down, right and left. When you code the program in BASIC, you have to change it to a one dimensional representation. That is, the control flow can only move up and down; from a low numbered line to a high numbered line, or visa versa.

Top down programming is a method which avoids this problem. Right from the beginning, your program is represented in a one dimensional list of instructions.

This section will explain how top down programming works, with a simple example to demonstrate. Later, a more complex example will be worked.

The first step, is to write down what the program is going to do. This should include what the display will look like, where this is applicable.

For example:
   This program will accept a number, N, then simulate throwing N dice. The total of the N dice will be printed.

of instructions which can be performed
consecutively. For the moment, don't worry if the
computer cannot perform the instructions. If you
can use BASIC words to describe what is to be
done, put them in: otherwise write it in English.

Example:
    1. INPUT N
    2. IF N=0 THEN STOP
    3. Throw a die N times, adding up the total
       as you go.
    4. PRINT total.

In this example, many of the steps are already in
BASIC, because it is a simple example. Usually,
most of the instructions will still be in English
at this stage. Next you have to break down the
instructions that are still in English even
further. This step is repeated until all
instructions are in BASIC.

The example, in instruction 3, calls for
something to be executed N times. That
immediately suggests a FOR loop. So now,
instruction 3 can be rewritten like this:

    3.1 FOR I=1 TO N
    3.2 throw a die and add to the total
    3.3 NEXT I

Now to break 3.2 down.

    3.2.1 LET DIE=INT(RND*5)+1
    3.2.2 LET SUM=SUM + DIE

You haven't finished yet. You need to check what
variables need to be initialized. There is one:-
SUM.

Now the program is finished. All that has to be

done is put the lines together, and renumber them.

```
100 LET S=0
110 INPUT N
120 IF N=0 THEN STOP
130 FOR I=1 TO N
140 LET D=INT(RND*5)+1
150 LET S=S+D
160 NEXT I
170 PRINT S
```

The next example is a game program called ACK-ACK. There is an aircraft flying across the screen, and you are in control of a guided missile, with which you try to hit the plane. A series of 10 planes makes up one game. At the end of the game, the number of planes that escaped your missiles is printed.

The plane will start at a random height, and move from right to left across the screen. The missile starts at the bottom in the middle of the screen, and moves up automatically. You have control of its movement to the left and right.

Since there are to be 10 planes, a FOR loop seems to be the best idea.

1. FOR J=1 TO 10
2. send a plane across the screen, and a missile up
3. NEXT J
4. PRINT the number that escaped.

Next the positions of the plane and missile need to be defined. MISSILE can be the horizontal position of the missile. It starts at 15, and moves when the player says. PLANE is the vertical position. It is randomly chosen, and does not alter. The other coordinates get smaller automatically, so by using a FOR loop, the control variable can be used as the other coordinate for both the plane and the missile. Before starting, the screen will need to be cleared of any old missiles or planes.

    2.1 LET MISSILE=15
    2.2 LET PLANE=INT(RND*10)
    2.3 CLS
    2.4 FOR I=20 TO 0 STEP -1
    2.5 print a plane, and blank out last position.
    2.6 print a missile, and blank last position.
    2.7 check if player wants to move missile, and work out next value of MISSILE.
    2.8 if the plane has been hit, print a message, increase the score, and GOTO 3
    2.9 NEXT I

Since the plane moves horizontally across the screen, by printing a blank after the plane, you will blank out the old position.

    2.5.1 PRINT AT PLANE,I;"▆▙▂▂▂▂▙-△"

Blanking out the old missile is a bit more difficult, since it can move in two positions. One way of solving this is to print the missile, and then straight away blank it out. This will cause it to blink (which can be an interesting effect), but you will still be able to see it. Another possibility is to clear the screen in each loop. For this example, we will use the first method. To slow the blinking a little, print the missile, then the plane, then blank out the missile. So get rid of 2.5.1, and use instead:-

    2.6.1 PRINT AT I,MISSILE;"▆▆"

2.6.2 PRINT AT PLANE,I;"▀╬▬▬▬█ -"
2.6.3 PRINT AT I,MISSILE;"▄▀▀"

2.7 requires the use of the INKEY$ function. The keys with the right and left arrows are a good choice of keys to move the missile right and left.

2.7.1 LET C=C+(INKEY$="8")-(INKEY$="5")

(Do you understand why this works? The computer evaluates (INKEY$="8") to 1 if it is true, and 0 if it is false. So an equivalent way of writing this would be

2.7.1 IF INKEY$="8" THEN LET C=C+1
2.7.2 IF INKEY$="5" THEN LET C=C-1

but the first way is shorter)

There are several ways of breaking up 2.8 One way would be:-

2.8.1 IF plane is hit THEN PRINT message
2.8.2 IF plane is hit THEN LET S=S+1
2.8.3 IF plane is hit THEN GOTO 3

But a better way, that only tests once, and therefore executes more quickly, is:-
2.8.1 IF plane is not hit THEN GOTO 2.9
2.8.2 PRINT message
2.8.3 LET S=S+1
2.8.4 GOTO 3

Now you have only to decide how to decide when the plane has been hit. The plane and the missile must be on the same line, so I=PLANE when the plane has been hit. This program only checks when the missile hits the target area of the plane (marked "■"), that is, MISSILE=I or I+1. Reverse these conditions to work out when the plane has

2.8.1 IF I<>PLANE OR (MISSILE<>I AND
      MISSILE<>I+1) THEN GOTO 2.9

(If you cannot work out how to reverse the
condition, you can put NOT in front of the
condition, as in

2.8.1 IF NOT(I=PLANE AND (MISSILE=I OR
      MISSILE=I+1)) THEN GOTO 2.9

The only variable that needs to be initialized is
S. It starts at 0.

ACK-ACK
c by Philip Thomas

```
100 LET S=0
110 FOR J=1 TO 10
120 LET M=15
130 LET P=INT(RND*10)
140 CLS
150 FOR I=20 TO 0 STEP -1
160 PRINT AT I,M;"▞▚"    (graphic T,Y)
170 PRINT AT P,I;"▟▀▀▀▀▙-▵" (graphic
    6,+,6,6,6,6,5; minus sign,space)
180 PRINT AT I,M;"  "
190 LET M=M+(INKEY$="8")-(INKEY$="5")
200 IF I<>P OR (M<>I+1 AND M<>I) THEN GOTO 250
210 PRINT AT P,I;"BOOOOM" (inverse)
220 PAUSE 100
230 LET S=S+1
240 GOTO 260
250 NEXT I
260 NEXT J
270 PRINT 10-S;" ESCAPED"
```

Now practise programming using a top down method.
It may seem a bit strange at first, but it is

much easier for larger programs than flowcharting. After a while, it will take less stages to break steps into smaller steps.

CHAPTER 18 **DEBUGGING**

You've just typed in your latest masterpiece of BASIC programming. All your friends have gathered round to see it work. You type RUN, and nothing happens! Or what did happen was not what you expected. What can you do? Well, apart from suggesting that you test your program before you invite you friends to watch it work, and lose your reputation as a programmer, this chapter will suggest several ways of getting your program working.

Programmers call errors in their programs 'bugs'. This sounds a lot less personal, and more as if it was someone (or something) else's fault than calling them 'mistakes'. (This piece of jargon might prove useful in saving your reputation.) Besides, bugs in programs are very elusive when you are trying to catch them, just like the real thing. The process of catching and eliminating bugs is called 'debugging'. And if it is any consolation, very few programs, apart from really simple ones, have no bugs at the beginning.

You will find your program a lot easier to debug if you have used a top-down method to design it, and have your notes in front of you. This will help you follow the logic of the program. There is no easy way to debug a program; it really is hard work. But the following methods will give you somewhere to start.

18.1 Catching the Bug

The first thing to do is work out what your program did do. Often, the program will stop with

a report code. This can be used to help you work
out what went wrong. The report code consists of
a number or letter, a slash, then another number.
The first number is the report code, and the
second is the line number.

Another possibility is that your program just
kept going and going and going. This is called an
infinite loop. If you press BREAK, the program
will stop, and you can find out where the
infinite loop is from the line number in the
report.

Alternatively, your program may have stopped
normally, but given the wrong answers. In this
case, you should try 'playing computer', or
'diagnostic statements'. These two techniques are
useful for finding out exactly why there is an
infinite loop too.

18.2 Report Codes

0 - This is the report you will get if your
program tries to execute a line that is numbered
higher than any existing line. For instance, if
you have the line GOTO 900, when the last line in
your program is line 200, the program will stop
with report 0. The line number will be the line
of the GOTO.

Usually, report 0 means successful completion of
the program. Unless the last line of the program
is a GOTO, a GOSUB, or a RUN statement, after
executing the last line, the computer will try to
execute the next line. If there is none, it
stops.

1 - You have used a variable in a NEXT statement
before it is set in a FOR statement. The computer

looked for a control variable with the name you gave, and could only find an simple variable.

The program:-

```
100 LET J=0
110 NEXT J
```

will cause error 1. You may have a GOTO, or GOSUB jumping to the middle of a FOR loop. Then the NEXT statement will be executed before the FOR statement, so check where your GOTO's and GOSUB's go.

2 - This report means that you have used an undefined variable. Check each variable in the line specified in the report code. A common cause of undefined variables is a typing error. The computer doesn't know that you really mean A1 when you type A.

For each simple variable, make sure it appears on the left hand side of a LET statement before it is used anywhere else. You cannot have a statement like

```
LET X = X+1
```

unless X has previously been assigned a value.

Also, the program

```
100 GOTO 120
110 LET X=0
120 LET X=X+1
```

is incorrect. Line 110 will not be executed before line 120 just because it is before it numerically.

All subscripted variables (except simple strings) must dimensioned in a DIM statement before they are used anywhere else.

All control variables must be set up in a FOR statement before the NEXT statement is executed.

Check all the variables, following the logic of the program. The methods described later in the chapter will help with this.

3 - This error applies only to subscripted variables. If the subscript is too large or too small, error 3 will result.

Note that in the ZX81, the first element in an array is always 1, not 0.

If you are using a subscripted variable in a FOR loop, check the highest and lowest values especially. Once again, you will need to follow the logic of the program, using one of the methods described later.

4 - This means that your program is too large, and the computer has used all its available memory. Correcting this problem is a suficiently large and important topic to deserve a chapter of its own. (See Chapter 19)

5 - This is the error message you get if there is not enough room on the screen for the print items. You may have tried to PRINT at a location that is not on the screen; line 22 for example.

If you repeatedly PRINT without clearing or scrolling the screen, you will eventually run out of room on the screen, even if you don't fill every line.

If this does not apply, then your program may have run out of memory. The computer tried to allocate more memory for the display file, but there was none. (See Chapter 19)

6 - This error is the result of 'arithmetic overflow'. Arithmetic overflow is the technical term for what happens when a number is too big for the computer to handle. The largest number the ZX81 can represent is approximately $10**38$.

To correct this you might be able to check the numbers in the calculation. For instance, with the EXP function, numbers over 83 will cause arithmetic overflow. So before calling EXP, you could check that its argument is not too big. For example, use

    100 IF X<=83 THEN PRINT EXP X

instead of

    100 PRINT EXP X

7 - Every RETURN statement must follow a corresponding GOSUB statement. If a RETURN is executed when there is no return address saved on the stack, error 7 will be the result. A possibility is that you have used a GOTO to jump to the subroutine instead of a GOSUB statement.

8 - This report code will never occur as the result of a bug in a program. It occurs if you try to use the INPUT statement as a command.

9 – This means that the computer has executed a STOP statement. This is not an error, so it needs no further explanation here.

A – If you use an invalid argument to a function, error A may result. For instance if you try to find the squareroot of a negative number using SQR. Instead, you could check the number before you call SQR, like this:-

    100 IF A>=0 THEN PRINT SQR A

rather than

    100 PRINT SQR A

B – If you use a number that is yoo large or too small as an argument to certain functions and statements, you may get error B. One of these functions is CHR$. The number must be between 0 and 255, so use

    100 IF X>=0 AND X<=255 THEN PRINT CHR$ X

instead of

    100 PRINT CHR$ X

To find out what range a function or statement will accept, check the reference manual entry for it.

C – This error relates only to the VAL function. The argument must be a string containing a valid numeric expression. Characters, unless they are the name of a numeric variable, are invalid. For example, "X" is a valid argument only if the variable X has already been defined in a LET statement.

D – This report is not caused by bugs. Either you pressed the BREAK key while the program was executing, or you started an INPUT line with STOP. Neither of these is an error.

E – The report E is not used.

F – If you use the command SAVE "", you will get this error. You must give a program name with a SAVE statement.

18.3 Pretend you are the computer

This is one way of working out what your program has done. Keeping your programming notes next to you, follow through your program, doing whatever the computer would do.

If there is a GOTO statement, then go to that line. As you work, write down the values of all the variables. If there are lots of GOTO statements in your program, jumping all over the place, you are going to have to be very careful not to get muddled.

If you have subroutines, write down the return address on your piece of paper, every time you get to a GOSUB statement. This is what the computer does, effectively. When you get to a RETURN statement, cross off the last return address.

If there is a loop in your program that is executed many times, you would be sitting there for years if you tried to work it through by

hand. Instead, you could get the computer to
help. Suppose you have a loop like this:-

```
100 LET A=0
110 FOR I=1 TO 100
120 LET A=A+I
130 NEXT I
```

Now, to check this loop, add these lines:-

```
125 PRINT I;"△ ";A;"△ ";
135 STOP
```

Type the command GOTO 100, and the computer will print the value of I and A after every iteration of the loop. It is imporatant that line 100 is executed before the loop, else A would be undefined. Instead, you could type the command LET A = 0, then GOTO 110. From this you can see a good reason to number your lines in 10's rather than consecutively, too.

For complicated computations, you could get the computer to do the arithmetic for you. For instance, suppose the program contains the line:

```
100 LET A = 2*PI*R
```

and you have worked out that the value of R at that stage is 12. Then you could type the command

```
PRINT 2*PI*12
```

and the computer would obediently print

75.396224.

18.4 Diagnostic Statements

These are statements you add too your program to

help you make a 'diagnosis' of what is wrong with
your program, like a doctor runs diagnostic tests
to find out what is wrong with a patient. These
may be PRINT statements that you add to find out
what the value of a certain variable is. Or you
could add STOP statements, then look at the
contents of any program variable, say X, by
typing PRINT X.

For example, the PRINT and STOP statements used
to test the FOR loop in the example on pretending
to be the computer, were diagnostic statements.
When you have finished debugging, the diagnotic
statements will be removed, of course.

All these methods help you find out what the
program is doing. This is not going to help
unless you have a clear idea of what it is
supposed to be doing. That is where your
programming notes can help. They should tell you
the logic behind the program design.

If you still cannot work out what is wrong with
the program, try explaining the program to a
friend that understands BASIC. Tell him how you
think it ought to work. He may be able to see an
error in the logic or programming because he is
looking at it differently. You may be able to
return the favour sometime.

18.5 Eliminating the Bug

Having found the bug (or bugs) by any or all of
the methods described, the next step is to get
rid of it. Finding bugs is usually the more
difficult of the two tasks, but eliminating them
is not trivial. If you make changes without
thinking about them, they are likely to have
unexpected side effects, and create more bugs.

The best policy is to go back and redesign the program.

Of course there are some bugs that are so simple, redesign is unnecessary. Mistyping a variable name is one example. All errors in program logic require redesign, of at least some part of the program (a subroutine, perhaps). Redesign is like design — it needs to be done on paper, not on the television screen.

After debugging a program or two, you will realize what a frustrating task it is. Well designed programs have fewer bugs, and are easier to debug. You will find that time spent in designing your programs reduces the time spent debugging.

18.6 Example

Here is an example of a program that has bugs in it. Type it into your computer, and follow the debugging by typing in the commands, and watching what happens.

Coin Tossing Game

In this game you start with $5. The computer will accept a $2 bet on either heads or tails, then tosses a coin. The game stops when you run out of money.

Program Design
```
          Assign the player $5
 TOSS     Ask "heads or tails"
          Input answer
          Toss a coin (0=Tails, 1=Heads)
          IF player loses THEN take $2 from  his
            money, and
                    IF he has no money THEN stop
                    ELSE goto TOSS
```

ELSE add $2 to his money and goto TOSS

Here is the first attempt at writing the program.
```
100 LET M=5
110 PRINT "HEADS OR TAILS?"
120 INPUT A$
130 LET C=INT (RND *2)
140 IF C=0 THEN LET C$="TAILS"
150 IF C=1 THEN LET C$="HEADS"
160 IF CODE A$=CODE C$ THEN GOTO 210
170 PRINT C$;" I WIN"
180 LET M=M-2
190 IF M=0 THEN STOP
200 GOTO 110
210 PRINT C$;" YOU WIN"
220 LET M=M+2
230 GOTO 110
```

Now RUN the program. For a while, everything seems fine, but then the program stops with error 5. If you look up what error 5 is, you will find that it means that there is no room on the screen. Since the program keeps writing to the screen without ever getting rid of anything, it soon fills the screen, and runs out of lines to write to.

Well that's easy to fix isn't it. Just add 105 CLS, and the program will start with a fresh screen each toss. No need to redesign this time! If you try that, you will discover that it doesn't work. The program returns to line 110 each time, so line 105 is not executed. Before redesigning to fix that error, let's check if there are any more bugs that can be fixed at the same time.

Let's check that the computer isn't cheating us out of any money. Insert a diagnostic PRINT statement —

```
115 PRINT M
```

Now RUN the program.

If you are unlucky enough to lose more often than you win, you will find that you have a negative ammount of money, but the game goes on. But the design says that if a player has no money, the game should stop.

To try and find out what is wrong, we will pretend to be the computer, and go through the program, starting with $1, because this is where the problem appears to be. So write down "M = 1"

Line 110 prints, and there is no problem there. Line 120 waits for input. Suppose "HEADS" is input. Write down "A$ = HEADS". Line 130 returns 0 or 1 randomly. (There is a problem in checking programs using RND in working out what will happen for all possibilities. In this case, the problem appears when the player loses, so try C = 0.) Write down "C = 0".

C is equal to 0, so write down C$ = "TAILS". C is not equal to 1, so do nothing for line 150. "H" is not equal to "T" so do nothing for line 160. Line 170 is a PRINT statement, so that is alright. Line 180 subtracts 2 from M, so cross out M = 1 and write M = -1. M is not equal to 0, so do nothing for line 190.

That is where the problem is! The program should have stopped there, because having a negative amount of money is worse than having no money at all.

There are a few other problems you may have noticed. It is possible to input something other than HEADS or TAILS. You will certainly lose if

you do, but it would be better if the program
treated this as a mistake.

It is very difficult to stop the program, other
than letting it stop when the screen becomes
full. You have to press NEWLINE, then BREAK very
quickly afterwards.

These are not really bugs, but redesigning the
progam is a good chance to include improvements.

Here is the redesigned program.

```
100 LET M=5
110 SCROLL
120 PRINT "MONEY LEFT ";M
130 SCROLL
140 PRINT "ANOTHER GO?"
150 INPUT A$
160 IF A$(1)="N" THEN STOP
170 SCROLL
180 PRINT "HEADS OR TAILS?"
190 INPUT A$
200 IF A$(1<>)"T" AND A$(1<>)"H" THEN GOTO 170
210 SCROLL
220 LET C=INT (RND*2)
230 IF C=0 THEN LET C$="TAILS"
240 IF C=1 THEN LET C$="HEADS"
250 IF CODE A$=CODE C$ THEN GOTO 270
260 PRINT C$;" I WIN"
270 LET M=M-2
280 IF M<=0 THEN STOP
290 GOTO 110
300 PRINT C$;" YOU WIN"
310 LET M=M+2
320 GOTO 110
```

CHAPTER 19 **SAVING MEMORY**

Since the ZX81 has only 1K of RAM, quite often you will find that your programs run out of memory. There a few methods you can use to reduce the amount of memory your program uses. There is a limit to how much you can compact you can make any program, so if after doing everything you can think of, you program still doesn't fit, you have two choices. You can get an additional memory pack, or you can give up on that program.

19.1 Compacting the code

The first way of compacting the program is to compact the code, while not affecting what it does. Here are a few examples of equivalent statements, where the first is longer than the second. You may have already thought of some of these, and after seeing a few examples, you may be able to think of even more.

Obviously, since REM statements have no effect, the first thing to go is any REM statements! It is better to have a program that is difficult to read, but works, than one that won't work because it runs out of memory.

1. Replace

    IF X=0 THEN ...

by

    IF NOT X THEN ...

2. Replace

by

    LET X=X-(X>3)

This works because X>3 is 1 if it is true, and 0 otherwise.

3. Replace

    100 IF INKEY$="5" THEN LET H=H+1
    110 IF INKEY$="6" THEN LET H=H-1

by

    100 LET H=H+(INKEY$="5")-(INKEY$="6")

This is similar to example 2, but saves a whole statement.

4. Use FOR loops if possible. For example

    100 LET I=0
    (program statements)
    150 LET I=I+1
    160 IF I<=N THEN GOTO 100

should be replaced with

    100 FOR I=0 TO N
    (program statements)
    150 NEXT I

5. If a subroutine is used only once in the program, put it directly into the main part of the program. That is, replace the GOSUB statement with the statement(s) in the subroutine, and delete the RETURN statement.

6. Make all variable names one character long.

7. Often in games, you will want to print a message and then STOP, if some condition is true. This would usually be written like this:-

    100 IF F<0 THEN PRINT "NO FUEL LEFT"
    110 IF F<0 THEN STOP

Instead of this you can make the program crash, by using an undefined variable in the first line, and save a line. You will get an error message when the program stops, but you can ignore it.

    100 IF F<0 THEN PRINT "NO FUEL LEFT";R

(R must be a variable that is not used anywhere else in the program).

19.2 Memory usage

If you need to save still more memory, then you will need to know how the ZX81 stores a program and its variables.

Each memory location is called a 'byte', and can hold a number between 0 and 255. This means it can contain the code for any single character.

The memory is divided into different areas, used for different purposes. Starting at location 16384, the system variables are stored. These are used as 'workspace' by the computer.

Your program is stored from location 16509 onwards. For each line of the program, there are 2 bytes for the line number, and 2 bytes to store the length of the line. Then follows the text of the line - one byte is used for each character you typed in, followed by the NEWLINE character, which takes one byte. Keywords, like PRINT and

INT, take only one byte.

Within the text, numerical constants (that is, numbers rather than variables) are not written as characters. Instead, they are written in binary form (which takes 5 bytes) preceded by the character code 126 to indicate that it is a number. This means that every time you use a number in the text of your statements, they use 6 bytes, no matter how large or small the number is. However, if you have a numeric variable with a one character name instead, it takes only one byte.

This opens a couple of possibilities for saving memory. The lines

    100 LET A=5
    110 LET B=5

could be replaced by

    100 LET A=5
    110 LET B=A

with a saving of 5 bytes.

All occurences of '0' can be replaced by PI-PI. (PI is a represented by a single character code, 66, since it is a keyword.) '1' can be replaced by PI/PI. Both of these save 3 bytes.

The next area in memory is for the display file. This gets bigger and smaller according to how much is to be printed on the screen. If your program has a large display, you could consider making it smaller to save memory, and use only the top left corner of the screen, rather than all of it. If the program is a game that uses graphics, it may not look as good with a smaller display, but it could mean the difference between a program that works, and one that doesn't.

Also, spaces are characters, and occupy room in the display file. However, when you clear the screen (CLS), all that is stored is NEWLINE characters. So overwriting whatever is on the screen with blanks is not the same as clearing the screen. If this is what your program does, you can save memory by using CLS instead.

The next area in memory is for storing program variables. Each numeric variable with a one character name takes 6 bytes:- one to store the name and 5 to store the number in binary form. For every extra character in the name, it takes an extra byte to store the variable.

If a number is used 5 times or more in the program, you can save space by defining a variable that is equal to it and using that variable wherever you would have used the number. Suppose the number '100' is used 5 times in your program. If you added the line

    100 LET S=100

that would take 14 bytes. (Can you see why?) It takes 6 bytes to store the variable, and one byte each time 'S' is used instead of '100'. That is a total of 25 bytes. On the other hand, if you used '100', it take 5*6 bytes, i.e. 30 bytes. That is a saving of 5 bytes.

19.3 Overlaying

The next technique for saving space is called overlaying, because you 'lay' one program 'over' another.

Try typing in the following program.

```
110 LET B=10
```

RUN the program, then delete the lines 100 and 110. Now type "PRINT A,B". The variables are still defined, even though the program lines are no longer in memory. This is the fact used in overlaying.

At the beginning of a program, you usually have variables to initialize. You can initialize these in a separate program, RUN the program, then carefully delete the lines of this first program. You cannot use NEW, since this would delete the variables you have just defined.

Now type in a second program containing the rest of the instructions. You cannot use RUN to start this program executing, because the variables stored by the first program would be deleted. Instead, you can use GOTO 1.

You can also initialize variables in the first program to replace numbers in the second program. This will save memory if there are two or more occurences of the number in the second program.

The following program illustrates many of the memory saving techniques described. It is written twice so that you can see what changes have been made. The first version will not fit in 1K, but the second does.

Before looking at the second version, try to compact the program yourself. You will learn to identify statements that can be compacted more quickly than if you just read the program through.

ESCAPE FROM THE DEATH STAR

c by Clifford Ramshaw

Princess Leda has been captured by Darth Vader, and is being kept prisoner on the death star. It is up to you to save her from his clutches, and bring her to safety.

When you start the game you will see a plan of the death star on the screen. The death star has two floors, connected by two lifts. Princess Leda (▙) is on the lower floor, and you are on the top floor. To save the princess, go over to her, and then she will follow you wherever you go. When you get back to your space ship (▟), the game stops.

The game also stops if you are caught by a storm trooper, but the end is not so nice for either you or the princess. You must avoid this at all costs. The storm trooper's movement is random, but weighted towards you.

The controls are 'Z' to move left, 'C' to move right, and 'M' to change floors. You can only change floors if you are at a lift.

The force be with you!

VERSION 1

```
100 LET W=-1      (W is your score)
110 LET W=W+1
120 LET Y=1       (Y is your vertical position)
130 LET X=3       (X is the storm trooper's vertical
                      position)
140 LET L=0       (Changes to 1 when princess is
                      saved)
150 LET S=15      (Your horizontal position)
160 LET H=15      (Storm trooper's horizontal
                      position)
170 GOSUB 100     (Print the screen)
```

```
180 PRINT AT X,S;"▲" (Blank out storm trooper)
190 PRINT AT Y,H;"△ " (Blank out your position)
200 IF H<18 AND INKEY$="C" THEN LET H=H+1
210 IF H>0 AND INKEY$="Z" THEN LET H=H-1
220 IF (H=1 OR H=11) AND INKEY$="M" THEN LET
    Y=Y+2
230 IF Y>3 THEN LET Y=1 (Calculate your new
                                position)
240 IF (X=Y AND RND>.5) OR RND>.7 AND S<H THEN
    LET S=S+1
250 IF (X=Y AND RND>.5) OR RND>.7 AND S>H THEN
    LET S=S-1
260 IF (S=1 OR S=11) AND RND>.5 THEN LET X=X+2
270 IF X>3 THEN LET X=1 (Calculate stormtrooper's
                                position)
280 PRINT AT X,S;"■" (graphic A - print
                                stormtrooper)
290 PRINT AT Y,H;"O"; (print your token)
300 IF L=1 THEN PRINT "▚" (graphic W - print the
                                princess if she is
                                following you)
310 IF S=H AND X=Y THEN PRINT AT 5,0;"SCORE ";W
320 IF S=H AND X=Y THEN STOP (the stormtrooper
                                got you!)
330 IF Y=3 AND H=18 THEN LET L=1 (you found the
                                princess)
340 IF H<>15 OR Y<>1 OR Y=0 THEN GOTO 180
350 PAUSE 50 (you have saved Princess Leia)
360 CLS
370 GOTO 110
1000 PRINT,"▆▆▆▆" (2*gr 7,gr R, gr space)
1010 PRINT,"▲▆▆▆" (space, gr T,2*gr space)
1020 PRINT "▆ ▆▆▆▆▆▆▆▆▆ ▆▆▆▆▆▆▆▆ " (gr 6, space,
                                9*gr 6, space, 8*gr 6)
1030 PRINT,"▲▲▲▚ "(3*space, gr W)
1040 PRINT "▆▆▆▆▆▆▆▆▆▆▆▆▆▆▆▆▆▆▆▆" (20*gr 6)
1050 RETURN
```

VERSION 2
```
100 LET W=W+T  (T is equal to 1 throughout the
```

program)
```
110 LET Y=T
120 LET X=3
130 LET L=PI-PI
140 LET S=15
150 LET H=S
160 PRINT,"▇▇"
170 PRINT,"▲▇"
180 PRINT "▇ ▇▇▇▇▇ ▇▇▇▇"
190 PRINT,"▲▲▲▇"
200 PRINT "▇▇▇▇▇▇▇▇▇▇▇▇"
210 PRINT AT X,S;"▲"
220 PRINT AT Y,H;"▲▲"
230 LET H=H+(INKEY$="C")-(INKEY$="Z")
240 IF (H=T OR H=11) AND INKEY$="M" THEN LET
    Y=Y+2*(Y=T)-2*(Y<>T)
250 IF (X=Y AND RND>.5) OR RND>.7 THEN LET
    S=S+(S<H)-(S>H)
260 IF (S=T OR S=11) AND RND>.5 THEN LET
    X=X+2*(X=T)-2*(X<>T)
270 PRINT AT X,S;"▇"
280 PRINT AT Y,H;"O";
290 IF L THEN PRINT "▇"
300 IF S=H AND X=Y THEN PRINT AT 5,PI-PI;
    "SCORE▲";W;Z
310 IF Y=3 AND H=18 THEN LET L=T
320 IF H<>15 OR Y<>T OR NOT L THEN GOTO 210
330 PAUSE 50
340 CLS
350 GOTO 100
```

Now, before you run the program, you will have to
type the commands
    LET W=-1 and LET T=1
since they are not defined in the program. To
start the program, type GOTO 1 rather than RUN,
so that these variables are not deleted.
Note that in line 300, Z is an undefined
variable. This is used to crash the program
rather than using an explicit STOP statement as
in the first version.

QUESTIONS

1. If the number 10 was used 7 times in a program, how many bytes would you save by replacing it with the variable Z?

2. How many times would you need to use the number 1 to make it worthwhile defining a variable S rather than using PI/PI?

3. How would you define an array of 10 numbers, 1 to 10 outside the main program? Define X to be that array, then use this program:-

        100 FOR I=1 TO 10
        110 PRINT X(I)
        120 NEXT I

to check that you have done it correctly. When you list the program, there should be only those 3 lines of code.

CHAPTER 20 **MACHINE CODE PROGRAMS**

20.1 Introduction

This chapter does not aim to teach you to write machine code, but will show you how to enter a machine code program into your ZX81.

The advantages of machine code are that it uses a lot less memory, and executes much faster than a BASIC program that does the same things. Against this, machine code is much more difficult to write, and requires considerable understanding of the way the computer works. BASIC is a compromise between machine code (which humans have trouble understanding) and English (which the computer cannot understand). The computer must first interpret what the BASIC statements mean, and this takes extra time. As you will see, machine code is very compact, so takes very little memory.

Machine code consists of 2 digit hex instuctions. (Hex is an abbreviation of hexadecimal, which refers to the base 16 number system). Hex uses the digits 0 to 9 and the letters A to F.

You need to place the hex instructions at a location that you know the address of, so that you can tell the computer where to start. You cannot put the instructions in the middle of a BASIC program. Try to type in anything but a BASIC statement, and the computer will not accept it.

One place you can put machine code is in a REM statement at the beginning of your program. Programs are always stored starting at location

16509. The first thing stored is the line number, which takes two bytes (or locations). Next the length of the line is stored, taking another two bytes. The REM takes another byte, so the contents of the REM statement will be stored starting from location 16514.

The REM is used to reserve the memory that will be used to store the instructions. You can put any characters you like in the REM; one character for each byte in the machine code program. Then you can use a BASIC program to POKE the machine code instructions into the REM statement.

Here is a BASIC program that does exactly that.

```
100 REM 11111111111111111111111111111111
        (32 1's)
110 LET S = 16514
120 FOR I=0 TO 31
130 SCROLL
140 GOSUB 500
150 PRINT I;"=";A$;" -> ";
160 INPUT A$
170 IF A$="" THEN GOTO 200
180 LET V=16* CODE A$+ CODE A$(2) - 476
190 POKE S+I,V
200 GOSUB 500
210 PRINT A$
220 NEXT I
230 STOP
500 LET V=PEEK (S+I)
510 LET H=INT (V/16)
520 LET L=V-16*H
530 LET A$=CHR$ (H+28) + CHR$ (L+28)
540 RETURN
```

If you RUN this program, the screen will display
    0 = 1D ->
and wait for input. The 1D is what is currently at the 0th position in the REM statement. 1D is

the hex code for the character '1', which is what
you put in the REM statement. You can change this
to anything you like, by inputting the hex code.

For instance, you could input the values

00, 01, 02, 03, 04, 05, 06, 07, 08, 09, 0A, 0B,
0C, 0D, 0E, 0F, 10, 11, 12, 13, 14, 15, 16, 17,
18, 19, 1A, 1B, 1C, 1D, 1E, 1F.

If you now LIST the program, the contents of the
REM statement has changed. It now contains the
first 32 characters in the character set.

20.2 Line Renumbering

We will now use the program to put something
useful into the REM statement. The following
machine code subroutine will renumber the line of
your program. This is useful if you are
developing a program and find that you want to
insert a line between two consecutively numbered
lines.

RUN the program again, entering the values:

11, A2, 40, 21, 64, 00, 1A, 3D, FE, 75, C0, 13,
1A, FE, 27, D0, 01, 0A, 00, 09, EB, 72, 23, 73,
23, 4E, 23, 46, 09, EB, 18, E6.

Now, to execute this subroutine, type the command

    PRINT USR 16514.

The USR function tells the computer to execute
the machine code subroutine beginning at location
16514. When it has finished, it will print
whatever is in the BC registers (a special
location in memory).

If you LIST the program you will find that not
only has the contents of the REM statement
changed again, but now the lines 500 to 540 have
been renumbered. Before you RUN the program
again, you will have to change lines 140 and 200
to GOSUB 240. This is because the program
renumbers the lines, but does not change the GOTO
and GOSUB statements.

Carefully delete lines 110 onwards, and type the
command CLEAR. You have, in line 100, a machine
code subroutine that will renumber the lines of
whatever program you have in memory. You can SAVE
it on cassette, and LOAD it before you develop a
program. Take care not to use line number 100 or
any number less than 100. The REM statement must
be the first in the program, else you won't know
where the subroutine begins. If you have a line
100 in your program, you will overwrite the
machine code.

20.3 Memory Left

Here is another useful machine code subroutine,
that will tell you how much memory you have left.

This is shorter than the line renumbering
subroutine. Type in the BASIC program, but this
time you will need only 13 1's in the REM
statement. Also, change line 120 to

    120 FOR I=0 TO 12

Enter these values

B7, ED, 5B, 1C, 40, ED, 62, 39, ED, 52, E5, C1, C9.

Now if you type

    PRINT USR 16514

the amount of memory left will be printed.

You can also use this subroutine within BASIC programs. For instance if you had the line

    150 IF USR 16514 < 50 THEN STOP

The program would STOP if there were less than 50 bytes of memory left.

QUESTION

Why wouldn't you use

    100 REM),inverse 6,RND,5,... etc.

rather than typing in a BASIC program to POKE 11,A2,40,21 etc into the REM statement? (11 is the character code of ')', A2 is the code of 'inverse 6', 40 is the code of 'RND', etc)

Hint:- Consider which character you would use for hex 64.

CHAPTER 21   SYSTEM VARIABLES

The ROM - the part of the memory you cannot alter - is really a program. It tells the computer what to do with your BASIC programs - how they are stored and executed, what is legal and what is not, how to perform mathematical calculations. Just like your programs, the 'system' needs variables to store information about the state of the computer.

These variables cannot be stored in ROM, because the values change and the contents of ROM cannot be changed. Bytes 16384 to 16508 - over 200 bytes of your 1K RAM - are reserved for use by the system. Some of these variables can be useful to you. Some give you information that can be used in programs. Others can be changed (using POKE) to make the computer do something that cannot be done in BASIC.

21.1 Using system variables.

Most system variables are 2 bytes long, so you have to PEEK at two locations to see what is in the variable. The location with the higher address contains the most significant part of the variable. Remember that a byte can hold a number up to 256, so the following expression will return the value of the variable.

        PEEK X + 256 * PEEK (X+1)

Poking a value V into X takes two statements.

        POKE X,V - 256 * INT (V/256)
        POKE X+1,INT (V/256)

21.2 The Variables

16388 and 16389 - RAMTOP
These addresses contain the address of the first byte after the BASIC system; that is, after your program, its variables, and all the stacks used during the execution of the program. Normally this is the last byte of the memory.

When the computer CLEARs the memory, it only checks as far as RAMTOP. As far as it is concerned, this is the last byte of the memory. If you POKE a lower value into RAMTOP, you will reserve a space that can be used for machine code routines. Then the routines will not be deleted by NEW, CLEAR or RUN.

CLS uses RAMTOP to determine how much memory there is. If there is more than 3.25K, then 1K is automatically reseved for the display file. If there is less memory than this, the display file will take up only as much memory as is absolutely necessary. After CLS has been executed, this is 25 bytes for 25 newline characters. CLS must be executed after the POKE for this to have any effect.

16398 and 16399   DF-CC
(Display File - Current Character)
These addresses contain the address in the display file of the current character. That is the character that is at the position that will be printed at next. This character is the one that will be overwritten when you next print something.

Here is a program, SCRAMBLE, that uses these locations. You have to fly your plane down one of the ditches, shoot the enemy base, and fly out of

the ditch and to safety to the right of the
screen. DF-CC is used to see if your plane is
about to crash into the wall of a ditch or the
enemy base. The controls are:-

    Q - up
    A - down
    M - forwards
    N - backwards
    V - fire

SCRAMBLE
C by Clifford Ramshaw

```
100 LET X=PI-PI
110 LET Y=X
120 LET H=X
130 PRINT
140 PRINT "▩▩▩▲▲▲▩▩▲▲▲▩▩▩▩"   (3*gr A,3 sp,2* gr A
                                3 sp,4*gr A)
150 PRINT "▩▩▩▲▲▲▲▩▲▲▲▲▩▩▩"   (3*gr A,4 sp, gr A
                                4 sp,3*gr A)
160 PRINT "▩▩▩▩▲▲▲▩▲▲▲▲▩▩▩"   (4*gr A,3 sp, gr A
                                4 sp,3*gr A)
170 PRINT "▩▩▩▩▲▲▲▩▲▲▲▲▲▩▩"   (4*gr A,3 sp, gr A
                                4 sp,3*gr A)
180 PRINT "▩▩▩▲▲▲▲▲▲▲▲▲▩▩▩"   (3*gr A,9 sp,3*gr A)
190 PRINT "▩▩▩▲▲▲▲▲▲▲T▲ ▩▩▩"  (3*gr A,6 sp,gr R,
                                gr E, sp, 3*gr A)
200 PRINT"▩▩▩▩▩▩▩▩▩▩▩▩▩▩▩"    (15 * gr A)
210 PRINT AT Y,X;"▲"
220 LET X=X+PI/PI + (INKEY$="M") - (INKEY$="N"
    AND X>PI-PI)*2
230 LET Y=Y+ (INKEY$="A") - (INKEY$="Q" AND
    Y>PI-PI)
240 IF X>VAL"14" AND H THEN PRINT "WELL DONE"
    (inverse);Q
250 IF X>VAL"14" THEN LET X=PI-PI
260 PRINT AT Y,X;
270 LET A$=CHR$ PEEK(PEEK 16398 + 256*PEEK 16399)
280 PRINT ">";
```

```
290 IF A$="■" THEN STOP
300 IF A$="▼" OR A$="▽" THEN GOTO 900
310 IF INKEY$<>"Y" THEN GOTO 210
320 PRINT "-----"
330 IF Y=6 AND X<9 THEN GOTO 900
340 CLS
350 GOTO 130
900 PRINT AT 6,9;"▲" (graphic Y,T)
910 LET H=PI/PI
920 IF X < VAL"9" THAN GOTO 210
```

You can also POKE DF-CC. This will cause the PRINT output to be stored at the location you POKE into it.

16418 - DF-SZ

Normally there are two lines at the bottom of the screen that you cannot print to. One is used to accept input, the other is blank. You can increase or decrease this number by poking the new number into DF-SZ.

If you use an INPUT statement whi;e there are no spare lines at the bottom of the screen, the system will crash. Remember to POKE 2 into DF-SZ before an input statement.

Try this program.

```
100 FOR I=1 TO 25
110 PRINT TAB 20;I
120 NEXT I
```

Now include this line

```
90 POKE 16418,0
```

and RUN the program again. (Note that poking this address in a command has no effect. You must have the statement in a program.)

16436 and 16437   -   FRAMES

This variable can be used for timing. In fact that is what PAUSE uses. Each time a frame is sent to the television, FRAMES is decremented by 1.

RAND 0 uses FRAMES to decide where RND should start in its sequence of pseudo random numbers.

16444 to 16476   -   PRBUFF

This is where the line that will be sent to the printer is stored. (16476 is a newline character.)

If you haven't a printer attatched, this area will not be used. USR routines can be stored here in that case, without taking any usable memory from your program. Routines stored here will be SAVEd and LOADed too.

As mentioned in chapter 19, memory is divided into different areas, which are used for different things. The systems variables area is the only area that remains the same size all the time. All other areas contain only enough room for the information they must contain. This means that their starting addresses change constantly. The computer keeps track of where each area begins and ends in variables in the system variables area.

16396 and 16397   -   D-FILE - contain the first address in the display file.

16400 and 16401 — VARS — is the beginning of the area that the program variables are stored in.

16404 and 16405 — E-LINE is the beginning of the area that is used to store the line that is currently being typed in or edited. There is also some workspace in this area.

There are also variables defining the stacks. The stacks grow and shrink during the execution of the program.

You cannot work out exactly how much memory you have free, except with a machine code program. However, using E-LINE, you can get an approximation.

    PRINT 17408-PEEK 16404 - 256*PEEK 16405

If this returns a value that is approximately 150, you are running out of memory. 150 bytes is approximately what is needed by the stacks.

CHAPTER 22   BEYOND GRAPHICS

One of the things that the program in ROM does is define the character set, that is what each character looks like. If you know how this is done, you can change the character set with a short machine code program. Unfortunately, you cannot define your own characters because the defintion must be in ROM, not RAM.

The system knows where to look for the character definition by looking at the contents of a register called the Interrupt Vector. (The book 'MACHINE PROGRAMMING MADE SIMPLE' explains the function of registers. Price and ordering details are at the back of this book.) A machine code program can change the contents of this register so that the characters change. These characters are more or less random, depending on what the contents of the ROM is at the location you tell the computer to look.

The assembly listing of a program to do this is:-

```
    LD A,n
    LD I,A
    RET
```

n is a number between 0 and 31. Every pair of numbers defines the same character set. That is 0 and 1 are the same character set. 30 and 31 define the ususal character set.

Here is a program which displays each character in each character set in turn. They are displayed very large so you can see them better. If you think the program is too slow, run it in fast mode.

```
100 FOR S=0 TO 15
110 LET A$="▞"   (graphic space)
120 LET B$="△"
130 FOR X=0 TO 63
140 PRINT "CHARACTER SET ";S
150 PRINT "CHARACTER ";X
160 PRINT "++++++++++"
170 FOR L=0 TO 7
180 LET V=PEEK(512*S+L+8*X)
190 LET P$="+"
200 LET D=256
210 FOR K=0 TO 7
220 LET D=D/2
230 LET C$=B$
240 IF V<D THEN GOTO 270
250 LET C$=A$
260 LET V=V-D
270 LET P$=P$+C$
280 NEXT K
290 PRINT P$+"+"
300 NEXT L
310 PRINT "++++++++++"
320 PAUSE 60
330 CLS
340 NEXT X
350 NEXT S
```

Now if you have decided which character set you want, you will need to use it in a program. Start your program with

```
    100 REM YOOOTAN
    110 POKE 16515,2*s
    120 IF USR 16514 THEN CLS
   9980 POKE 16515,30
   9990 IF USR 16514 THEN CLS
```

(s is the number of the character set, as displayed in the previous program. TAN is the function, not three characters.)

Now type these commands

    POKE 16516,237
    POKE 16517,71

The REM statement now contains the machine code program to change the contents of the Interrupt Vector.

Your program should go between the lines 120 and 9980. Remember that you cannot write messages when you are using an alternative character set — the message will be totally unintelligable.

You can use any lines that call USR 16514 instead of lines 120 and 9990

Here is a program which uses character set 5. You are a marathon runner in an obstacle race along a beach. Below you is the sea, and along your path are many holes. If you fall into a hole, or run into the sea, then you are out of the race.

The controls are:-
    M - up
    Z - down

```
100 REM YOOOTAN
110 POKE 16515,10
120 IF USR 16514 THEN CLS
130 FOR X=PI-PI TO 30
140 FOR Y=PI-PI TO PI+INT(PI*RND)
150 PRINT AT Y,X;" " (graphic space)
160 NEXT Y
170 FOR Y=Y TO 6
180 PRINT AT Y,X;"P"
190 NEXT Y
200 NEXT X
210 FOR X=PI/PI TO 15
```

```
220 PRINT AT INT(5*RND),PI+(27*RND);"U"
230 NEXT X
240 LET Y=PI/PI
250 LET X=-Y
260 PRINT AT Y,X;" "  (graphic space)
270 LET Y=Y-(INKEY$="M")+(INKEY$="Z")
280 LET Y=Y*(Y)PI-PI)
290 LET X=X+PI/PI
300 PRINT AT Y,X;" "
310 IF X=30 THEN RUN
320 IF PEEK(PEEK 16398 + 256*PEEK 16399)=128 THEN
    GOTO 260
330 POKE 16515,30
340 IF USR 16514 THEN CLS
```

Before RUNning this program, type in the commands:-
    POKE 16516,237
    POKE 16517,71

Also note that spaces have been redefined. If you PRINT AT 10,8 then the character you print will be preceded by 7 of character 0 in the character set you are using.

QUESTIONS

1. Which character set is used if you type
    POKE 16515,31
    RAND USR 16514

2. Look for another character set, and replace the characters in the program above.

APPENDIX A CHARACTER SET

The following is a table which lists the characters together with their codes :-

CODE	CHARACTER		
0	space	32	4
1	▰	33	5
2	▰	34	6
3	▰	35	7
4	▰	36	8
5	▰	37	9
6	▰	38	A
7	▰	39	B
8	▰	40	C
9	▰	41	D
10	▰	42	E
11	"	43	F
12		44	G
13	$	45	H
14	:	46	I
15	?	47	J
16	(	48	K
17	)	49	L
18	>	50	M
19	<	51	N
20	=	52	O
21	+	53	P
22	-	54	Q
23	*	55	R
24	/	56	S
25	;	57	T
26	,	58	U
27	.	59	V
28	0	60	W
29	1	61	X
30	2	62	Y
31	3	63	Z
		64	RND
		65	INKEY$
		66	PI

NOTE : THE CODE BETWEEN 67 TO 111 ARE NOT USED..........

112	cursor up
113	cursor down
114	cursor left

115	cursor right	154	inverse ,
116	GRAPHICS	155	inverse .
117	EDIT	156	inverse 0
118	NEWLINE	157	inverse 1
119	RUBOUT	158	inverse 2
120	K / L mode	159	inverse 3
		160	inverse 4
121	FUNCTION	161	inverse 5
122	not used	162	inverse 6
123	not used	163	inverse 7
124	not used	164	inverse 8
125	not used	165	inverse 9
126	number	166	inverse A
127	cursor	167	inverse B
128	■	168	inverse C
129	■	169	inverse D
130	■	170	inverse E
131	■	171	inverse F
132	■	172	inverse G
133	■	173	inverse H
134	■	174	inverse I
135	■	175	inverse J
136	■	176	inverse K
137	■	177	inverse L
138	■	178	inverse M
139	inverse "	179	inverse N
140	inverse	180	inverse O
140	inverse	181	inverse P
141	inverse $	182	inverse Q
142	inverse :	183	inverse R
143	inverse ?	184	inverse S
144	inverse (	185	inverse T
145	inverse)	186	inverse U
146	inverse >	187	inverse V
147	inverse <	188	inverse W
148	inverse =	189	inverse X
149	inverse +	190	inverse Y
150	inverse -	191	inverse Z
151	inverse *	192	''''
152	inverse /	193	AT
153	inverse ;	194	TAB

195	not used	236	GOTO
196	CODE	237	GOSUB
197	VAL	238	INPUT
198	LEN	239	LOAD
199	SIN	240	LIST
200	COS	241	LET
201	TAN	242	PAUSE
202	ASN	243	NEXT
203	ACS	244	POKE
204	ATN	245	PRINT
205	LN	246	PLOT
206	EXP	247	RUN
207	INT	248	SAVE
208	SQR	249	RAND
209	SGN	250	IF
210	ABS	251	CLS
211	PEEK	252	UNPLOT
212	USR	253	CLEAR
213	STR$	254	RETURN
214	CHR$	255	COPY
215	NOT		
216	**		
217	OR		
218	AND		
219	<=		
220	>=		
221	<>		
222	THEN		
223	TO		
224	STEP		
225	LPRINT		
226	LLIST		
227	STOP		
228	SLOW		
229	FAST		
230	NEW		
231	SCROLL		
232	CONT		
233	DIM		
234	REM		
235	FOR		

APPENDIX 2 OPERATOR PRIORITY

OPERATOR	PRIORITY
Subscript, slicing	12
All functions	11
**	10
− (negative sign)	9
*, /	8
+, −	6
=, ⟩, ⟨, ⟨⟩, ⟨=, ⟩=	5
NOT	4
AND	3
OR	2

APPENDIX C **APPLICATIONS**

Financial - Cheque book balancing

This program will help you keep your cheque book in order.

There are 3 transactions that the program will process. When the computer prints "TRANSACTION?" you can type

    C to write a cheque
    D to deposit to your cheque account
    S to reconcile a statement with the cheques written.
    X to exit from the program.

Now each transaction will be explained more fully.

C. You will be asked "AMOUNT?". Type in the amount of the cheque, and the computer will display information about your account which should be recorded on the butt of the cheque:- balance brought forward, deposits since the last cheque, the amount of this cheque and the resulting balance.

Next you will be asked "NO?". Type in the number of the cheque - a number between 1 and 24 that will identify the cheque. Don't use a number that you have used since the last statement reconciliation, nor a number that was an unpresented cheque in the last statement reconciliation.

D. You will be asked "DEP. AMT?". Type in the amount deposited to the account.

S. First you will be asked "NUMBER?". Type in the number of the first cheque on the statement. Next you will be asked "AMOUNT?", so type in the amount of that cheque. If there are more cheques, type "Y" in response to "MORE?" and repeat the process of entering cheque number and amount for each cheque.

When there are no more cheques, enter "N" and the computer will print a list of unpresented cheques, together with their total.

Bank Charges — enter any bank charges, interest etc as a cheque, but give the number as 0. 0 is not a legal cheque number because it is used exclusively for this purpose.

You will need to do a statement reconciliation every 24 cheques at least, or you will run out of numbers for the cheques. This is because of limitations of the size of the ZX81's memory. If you have more memory, you can make the following changes to allow 99 cheques.

```
140 DIM A$(200)
5710 FOR X=PI/PI TO 199 STEP Z
```

Also you could improve the prompts. The program does no error checking. You could print a warning if you write a cheque that leaves a negative balance. You could check for repeated cheque numbers.

```
100 LET Y=10
110 LET Z=2
120 LET B=Z-Z
130 LET T=B
```

```
140 DIM A$(40)
150 LET D=B-B
160 CLS
170 PRINT "TRANSACTION?"
180 INPUT X$
190 CLS
200 GOTO CODE X$*Y*Y

4000 PRINT "AMOUNT?"
4010 INPUT X
4020 CLS
4030 PRINT "B.F.▲";B,"DEP▲";D,"CHEQUE▲";X
4040 LET B=B+D-X
4050 PRINT "BAL▲";B,"NO.?"
4060 INPUT N
4070 CLS
4080 IF N THEN LET A$(Z*N+PI/PI)=STR$ INT(N/Y)
4085 IF N THEN LET A$(Z*N+Z)=STR$(N-INT(N/Y)*Y)
4090 IF N THEN LET T=T+X
4095 GOTO 150

4100 PRINT "DEP AMT?"
4110 INPUT X
4120 LET D=D+X
4130 GOTO 160

5600 CLS
5610 PRINT "NUMBER?"
5620 INPUT X
5630 LET A$(Z*X+PI/PI TO Z*X+Z)="△△"
5640 PRINT "AMOUNT?"
5650 INPUT X
5660 LET T=T-X
5670 PRINT "MORE?"
5680 INPUT X$
5690 IF X$="Y" THEN GOTO 5600
5700 CLS
5710 FOR X=PI/PI TO 39 STEP Z
5720 IF A$(X)<>"△" THEN PRINT A$(X TO X+PI/PI);
     "△";
5730 NEXT X
```

```
5740 PRINT "TOTAL▲";T
5750 GOTO 170
```

Line 200 contains an expression you may not have met before. It is called a 'computed GOTO' – meaning that the line number has to be computed before the GOTO can be executed. So the program does not always jump to the same line.

CODE X$ returns the character code of the first letter in the string you input. If you input "D" this is 41. Y is used to save memory. It is always 10. So when you input "D", the program jumps to line 4100.

Educational - Geometry test

This program will ask questions related to geometry - what is the circumference of a circle of radius 3? - for example. It will accept any number within .5 of the correct answer as right. Note that you do not have to calcuate the answer; you can enter an arithmetic expression.

The program is in two parts. The first part sets up the vocabulary of the program. A$ contains the words used. Sentences are defined by the starting position of each word. These positions are stored in the REM statement at line 100. A$(56) is the first letter of 'WHAT'. A$(61) is the first letter of 'IS', and so on. Each sentence contains 9 words.

```
100 REM 123456789012345678901234567890123456
        (36 characters)
110 FOR I=0 TO 35
120 SCROLL
130 PRINT I;"▲->▲";
140 INPUT J
150 POKE 16514+I,J
160 NEXT I
170 LET A$="SURFACE▲AREA▲CIRCUMFERENCE▲VOLUME▲
        CIRCLE▲SPHERE▲RADIUS▲WHAT▲IS▲THE▲O
        F▲WITH▲"
```

RUN this program and enter the following numbers

56, 61, 64, 14, 68, 12, 35, 71, 49
56, 61, 64, 9, 68, 12, 35, 71, 49
56, 61, 64, 1, 9, 68, 12, 42, 49
56, 61, 64, 28, 68, 12, 42, 71, 49

Now delete lines 110 onwards, and enter this program.

```
110 CLS
120 LET Q=INT (RND*4) + PI/PI
130 LET M=INT (RND*20) +2
140 GOSUB Q*1000
150 FOR I=Q-Q TO 8
160 LET J=PEEK (16505 +Q*9 + I)
170 PRINT A$(J);
180 IF A$(J)="▲" THEN GOTO 210
190 LET J=J+PI/PI
200 GOTO 170
210 NEXT I
220 PRINT M;"?"
230 INPUT R
240 IF ABS(R-A)>.5 THEN GOTO 270
250 PRINT "YES, ";A
260 GOTO 280
270 PRINT "NO, ";A
280 PRINT "PRESS N/L WHEN READY"
290 INPUT X$
300 GOTO 110

1000 LET A=2*PI*M
1010 RETURN

2000 LET A=PI*M*M
2010 RETURN

3000 LET A=4/3*PI*M*M
3010 RETURN

4000 LET A=PI*M*M*M/3
4010 RETURN
```

Don't use RUN to start the program. This would delete A$. Use GOTO 1 instead.

The questions in the program could be replaced by any others by redefining the vocabulary and sentences. The subroutines work out the answers to the questions. If you have more memory, you could have more than 4 questions.

Games — Hot air balloon

This is a graphics game. You must try to blow the hot air balloon to the top of the screen. You do this by keeping your token (at the bottom of the screen) directly underneath the balloon.
As your score increases, the difficulty of blowing the balloon upwards increases too.

```
100 PRINT "DIFFICULTY? (>=0)"
110 INPUT G
120 LET A=16
130 LET S=0
140 LET D=32
150 LET F=(S+G)*20
160 IF F<3 THEN LET F=3
170 FOR H=40 TO 4 STEP -1
180 CLS
190 PRINT AT 20,A;" " (inverse +)
200 LET A=A + (A<31)*(INKEY$="8") -
    (INKEY$="5")*(A>0)
210 PLOT D,H
220 LET D=D+(RND*5)-2+2*((D<2)-(D>60))
230 IF F THEN LET F=F-1
240 IF A*2<D-3 OR A*2>D+3 THEN GOTO 280
250 IF NOT F THEN LET H=H+1
260 IF H>40 THEN GOTO 300
270 LET H=H+1
280 NEXT H
290 LET S=S-2
300 LET S=S+1
310 PRINT
320 PRINT "SCORE▲";S
330 PAUSE 200
340 GOTO 140
```

Games — Galaxian

This is a 1K version of the arcade game.

There is a fleet of enemy fighters above you. Suddenly, one dives at you, guns blazing! Can you shoot him before he shoots you or crashes into you? You can only shoot the fighter that is diving at you. If you miss him, he will come back and dive again.

The game ends when you have shot the entire fleet, or you are hit.

The controls are:-

    Z — left
    C — right
    M — fire


```
100 LET V=10
110 LET S=PI/PI
120 LET G=5
130 LET X=4+S*3
140 LET Y=PI-PI
150 PRINT TAB X;
160 FOR I=PI/PI TO 5-S
170 PRINT "♣△"; (gr T,gr 4,space)
180 NEXT I
190 LET G=G+(INKEY$="C") - (INKEY$="Z")*(G>PI/PI)
200 PRINT AT V,G;"△♣△" (sp, gr Q,gr 4,sp);
    AT Y,X;"△△"
210 LET X=X+(Y>3)-(Y<4)
220 LET Y=Y+PI/PI
230 PRINT AT Y,X;"♥" (gr Y,gr 1)
240 IF Y=V THEN GOTO 360
250 IF RND<.8 THEN GOTO 300
260 FOR I=Y+PI/PI TO V
270 PRINT AT I,X;"■" (gr 8);AT I,X;"△"
280 NEXT I
```

```
290 IF X=G THEN GOTO 380
300 IF INKEY$<>"M" THEN GOTO 190
310 FOR I=V-PI/PI TO PI/PI STEP -PI/PI
320 PRINT AT I,G;"▮" (gr 8);AT I,G;"▵"
330 NEXT I
340 IF G<>X AND G<>X+PI/PI THEN GOTO 190
350 LET S=S+PI/PI
360 CLS
370 IF S<>5 AND (ABS(X-G)>PI/PI OR Y<>V) THEN
    GOTO 130
380 PRINT S
390 PAUSE 50
400 CLS
410 RUN
```

Games — Space Rendezvous

You are in command of the small space craft on the launchin pad. You must refuel your ship in midspace — a difficult manouver that only the most experienced commanders would dare attempt.

Enter the gravity value for your location (between 0.1 and 0.6 is best — 0.6 is VERY difficult). As you move, your fuel supply diminishes. Watch the fuel indicator below the ground.

Controls are:-
    X — down
    W — up
    D — right
    A — left

```
100 LET X=PI/PI
110 LET Y=11
120 LET A=X
130 INPUT G
140 PRINT AT 12,PI-PI;"▬▬▬▬▬▬▬▬▬▬"
    (gr W,gr 6,gr Q,14*gr A)
150 PRINT "▬▬▬▬▬▬▬▬▬▬" (15*gr 6)
160 LET F=58
170 PRINT AT Y,X;"▲"
180 LET Y=Y+G+(INKEY$="X")-(INKEY$="W" AND
    Y>PI/PI)
190 IF Y>11 THEN LET Y=11
200 LET X=X+(INKEY$="D" AND X<15)-(INKEY$="A"
    AND X>PI-PI)
210 LET F=F-(INKEY$<>"")
220 PRINT AT Y,X;">";AT PI-PI,A;"▲▲"
230 LET A=A+VAL".5"
240 IF A>15 THEN LET A=PI-PI
250 PRINT AT PI-PI,A;"▬▬" (gr W,gr 4)
260 UNPLOT F/2,16
270 IF F<=PI-PI THEN GOTO 500
```

```
280 IF INT(Y+.5)<>PI-PI THE GOTO 170
290 IF X=A OR X=A+PI/PI THEN GOTO 500
300 IF X+PI/PI<>A THEN GOTO 170
310 PRINT,,"WELL DONE","SCORE: ";G*10*F
320 PRINT,,,,"CRASHED" (inverse)
```

Artificial Intelligence - Gomoku

This program demonstrates how the computer can be used to perform the same functions as the human mind. That is, the computer can appear to be intelligent.

This program plays the game Gomoku. This game is played on a 19 x 19 grid usually. Each player takes turns to put a piece on the grid. The aim is to get 5 pieces in a row - horizontally, vertically or diagonally.

This program uses an 8 x 8 grid, and plays defensively. The computer checks which is the longest string that your piece forms, and blocks that string.

```
 10 REM■■■" (gr S,gr D,gr 1,)
 20 FOR I=PI/PI TO 8
 30 PRINT AT I,I-I;I;"........" (8 .'s); AT I-I,I;I
 40 NEXT I

100 INPUT M
110 LET X=M
120 GOSUB 900
130 IF X$<>"." THEN GOTO H*H
140 PRINT "▪"   (inverse O)

300 LET B=PI-PI
310 FOR I=B TO PI
320 LET C=B-B
330 LET X=M
340 LET O=PEEK(16514+I)
350 LET X=X-D
360 GOSUB 900
370 IF X$<>" " (inv O) AND D<B-B THEN GOTO 420
380 LET C=C+(D<B-B)
390 IF X$="." AND D>B-B THEN LET R=X
```

```
400 IF X$<>" " (inv 0) AND D>B-B THEN LET D=-D
410 GOTO 350
420 IF C<B THEN GOTO 460
430 LET B=C
440 LET J=X
450 IF X$<>"." THEN LET J=R
460 NEXT I
470 IF B>I THEN STOP

500 LET X=J
510 GOSUB 900
520 IF X$<>"." THEN GOTO 550
530 PRINT "X"
540 GOTO H*H
550 X=INT(RND*H*H)
560 GOTO 510

900 PRINT AT INT(X/H),X-H*INT(X/H);
910 LET X$=CHR$ PEEK(PEEK 16398+256*PEEK 16399)
920 RETURN
```

If you RUN this program you will find undefined variables. So type in:-

```
    LET H=10
    LET R=0
```

Now type GOTO 1 to start the program.

If you have more than 1K, then you can add a few extra lines to inmprove the program.

```
    160 LET H=10
    170 LET R=0
```

These lines will stop the undefined variables when you RUN the program.

```
        470 IF B=PI/PI THEN GOTO 550
```

This will make the program more friendly. Instead of blocking every move you make, if you put down a piece that has no neigbours, it will make a random move.

Other changes could be made. You could print a message when the game finishes.

If you have pieces like this:-
```
     O OOO
```
the computer may add
```
     O OOOX
```
leaving you a winning move. You could add more checks to prevent this. You could make the computer play a more aggressive game too. It does not check if there is a winning move it can make at the moment either.

REFERENCE MANUAL

Your ZX81 recognizes many 'keywords'; words which have a special meaning in BASIC. There are two kinds of keywords:- functions and statements.

STATEMENTS can be used either as commands or in programs. (INPUT is the only exception; it can only be used in a program.) Statements tell the computer to do something - PRINT something or RUN a program, for instance.

FUNCTIONS cannot be used by themselves. A function must be part of a statement. For instance, SQR 9 calculates the squareroot of 9. It returns the value 3, and you can use SQR 9 wherever you would use 3. Just as it would not make sense to type just '3' into the computer, it does not make sense to type SQR 9. The computer would not know what you meant it to do with the 3.

The 9 in SQR 9 is called an argument or operand. Many functions require arguments. They may be numbers or strings, depending on the function.

In this section, all functions and statements recognized by your ZX81 are explained. They are arranged in alphabetical order, to make a complete, easy-to-look-up ZX81 BASIC Reference Manual.

ABS

This function must be given a number. Its value is the value of the number for a positive number, or the negative of the number if the number is negative.

For example ABS -2 is equal to 2; ABS 2 is equal to 2; and ABS 0 is equal to 0. To see what ABS does , run this program.

```
100 PRINT "X","ABS X"
110 FOR I=-5 TO 5
120 PRINT I,ABS I
130 NEXT I
```

AND

There are three different ways of using AND:

1. It can be used in the condition of an IF...THEN statement, like this:-

>   IF DATE=25 AND MONTH$="DECEMBER"
>   THEN PRINT "MERRY CHRISTMAS"

Both the conditions must be true before the message will be printed. In other words, the function AND in this case implies a repetition of the IF - you could think of it as

>   IF TODAY IS THE 25th
> AND IF IT'S DECEMBER
> THEN IT MUST BE CHRISTMAS.

but in programming you only need the one 'IF' statement.
As many AND statements (and OR statements) can be linked together in one IF ... THEN line as desired.

If neither of the conditions is true, or just one of the conditions is true then MERRY CHRISTMAS will not be printed.

example program:
```
100 INPUT X
110 IF X>=0 AND X<10 AND X = INT X THEN GOTO 140
120 PRINT"THAT WAS NOT A DIGIT"
130 RUN
140 PRINT "THAT WAS A DIGIT"
150 RUN
```

In order to qualify as a digit, the number you

input has to fulfill three conditions:- it has to be greater than or equal to zero, less than ten, and an integer number. Try a few different numbers and see that this is what the AND does.

2. AND can also be used as a numeric operator, as in the following statement.

    LET A = 2 AND B

This is equivalent to these two statements
    LET A = 2
    IF B = 0 THEN LET A = 0

That is, if the second operand (in this case B) is zero, then A is assigned zero, otherwise it is assigned the value of the first operand (in this case 2).

This is because the AND treats the second number as a logical value being TRUE or FALSE - whether the number is TRUE or FALSE depends solely on whether the number are non-zero (TRUE) or zero (FALSE).

3. The AND statement can be used to link a string and a number (with the number being used as a logical value of TRUE or FALSE).

The first operand must be a string. This use of AND works in the same way as the second use of AND described above.

e.g.   LET A$ = B$ AND C

The second operand must be a number, or numeric variable, which will determine whether it is TRUE that A$ = B$, or FALSE that A$ = B$ (in which case A$ will be equal to "").

A$ will be the null string if C is zero; any

other value of C will make A$ be equal to B$

example program:
```
100 PRINT "WHAT IS YOUR NAME?"
110 INPUT B$
120 PRINT "WHAT IS YOUR AGE?"
130 INPUT C
140 LET A$ = B$ AND C
150 PRINT "HELLO ";A$;"."
```

If you input your age as zero, then your name will not be printed, (because A$ is the null string) otherwise your name will be printed.

ARCCOS (ACS)

The ARCCOS function computes the angle (A) between the hypotenuse (H) and one side (X) of a right angle triangle. Its argument is the ratio of the lengths of these sides.

$$ARCCOS(X/H) = A$$

The angle is given in radians. A radian is approximately 57 degrees. (To be exact, 180 degrees = PI radians).

The opposite of ARCCOS is COS, so
    COS A = X/H
Since COS always returns a value between -1 and 1 (because the hypotenuse is longer than or equal to the other side) the ARCCOS of a value not in this range is undefined. You will get error A if the number is not a valid argument.

This program given will plot the ARCCOS function.
```
100 PRINT AT 6,0;"3"
110 PRINT AT 11,0;"2"
120 PRINT AT 16,0;"1"
130 PRINT AT 21,1;"-1"
140 PRINT AT 21,11;"0"
150 PRINT AT 21,21;"1"
160 LET X=-1
170 FOR Y=3 TO 42
180 PLOT Y,INT(ARCCOS(X)*10)
190 LET X=X+.05
200 NEXT Y
```

ARCSIN (ASN)

The ARCSIN function computes an angle (A) in a right angle triangle. Its argument is the ratio of the lengths of its hypotenuse (H) and the side opposite the angle (Y)

    A = ARCSIN (Y/H)

The angle is given in radians. A radian is approximately 57 degrees. (To be exact 180 degrees = PI radians).

The opposite of ARCSIN is SIN.
    SIN A = Y/H
The ratio Y/H must be between −1 and 1 since the hypotenuse of a right angle triangle is longer than or the same length as the other sides. If you give the ARCSIN function an argument outside this range, you will get error A.

This program will plot the ARCSIN function
```
100 PRINT AT 6,0;"3"
110 PRINT AT 11,0;"2"
120 PRINT AT 16,0;"1"
130 PRINT AT 21,1;"-1"
140 PRINT AT 21,11;"0"
150 PRINT AT 21,21;"1"
160 LET X=-1
170 FOR Y=3 TO 42
180 PLOT Y,INT(ARCSIN(X) * 10)
190 LET X=X+.05
200 NEXT Y
```

ARCTAN (ATN)

The ARCTAN function computes an angle in a right angle triangle. Its argument is the ratio of two sides, neither of them the hypotenuse. The ratio is given adjacent side (X) to the opposite side (Y).

    A = ARCTAN (X/Y)

The angle is given in radians. (180 degrees equal PI radians.)

This program will plot the ARCTAN function.
```
100 PRINT AT 16,0;"-1"
110 PRINT AT 11,0;"0"
120 PRINT AT 6,0;"1"
130 PRINT AT 21,2;"-2"
140 PRINT AT 21,12;"0"
150 PRINT AT 21,22;"2"
160 LET Y =-2
170 FOR X = 4 TO 50
180 PLOT X,INT (ARCTAN(Y)*10 +20)
190 LET Y=Y+.1
200 NEXT X
```

BREAK

The BREAK key (on the same key as SPACE) will interrupt the program that is running. It will return the report code D, and the line at which it was interrupted.

The BREAK is not recognized in some situations. If the computer is waiting for input from the keyboard, the key will be recognized as a SPACE.

To continue running from where the program stopped, use the CONT command (see CONT).

CHR$

The computer stores characters as numbers. It has a code which it uses to convert from one to the other. The CHR$ function enables you to find out what character corresponds to a code number.

Code numbers are between 0 and 255. If you call the funtion with an argument outside this range, you will get error B.

The following program will print all the characters and their corresponding codes.

```
100 FOR X=0 TO 255
110 PRINT AT 21,0;X;"△";CHR$ X
120 SCROLL
130 NEXT X
```

You will notice that next to quite a few numbers there are question marks. 15 is the code for '?'. Some of these are numbers that are not used, the others correspond to unprintable characters, like the newline character for instance.

Key words also have code numbers. For instance, the key word PRINT has code number 245.

CLEAR

The CLEAR statement removes all variables, and frees the space they occupied. All variables become undefined.
Try this to see what clear does. Type in:-

100 LET A$="HELLO THERE"

Now RUN the program. If you now type PRINT A$, the message "HELLO THERE" will appear.

Now type CLEAR, then PRINT A$. This time you will get an error message. The error code 2 means that an undefined variable has been used.

Your program is still there though. Just RUN it again, and A$ will be in memory once again.

CLS

The CLS statement clears the screen, and the display file. Your program and its variables are not effected.

```
example program
100 PRINT "GOING..."
110 PAUSE 50
120 PRINT "GOING..."
130 PAUSE 50
140 PRINT "GONE"
150 PAUSE 10
160 CLS
```

Because this command clears the display file, it frees the memory that was occupied by the file. This is useful if you have a lot printed on the screen, and your program is running out of memory.

CODE

This function is given a string as its argument. The value of CODE is the character code of the first character in the string. If the string is empty ("" is the empty string) then CODE is zero.

For example, CODE "STRING" is equal to 56, because the code of 'S' is 56. CODE "STOP" is also 56, and so is any other word beginning with 'S'.

```
example program
100 LET X = 1
110 FOR Y=1 TO 20
120 PRINT X
130 LET X=X+1
140 NEXT Y
150 PRINT "DO YOU WANT ME TO COUNT HIGHER?"
160 INPUT A$
170 IF CODE A$<>62 THEN STOP
180 CLS
190 GOTO 110
```

If you type in "YES" or "Y" or anything else beginning with "Y", the program will keep going; otherwise it will stop.

COS

If you give the COS function an angle (in radians) it will calculate the ratio between the hypotenuse (H) and the side adjacent to the angle (X) for a right angle triangle.

$$COS\ A = (X/H)$$

The number returned will be between -1 and 1, because the hypotenuse will be at least as long as the other side.

This program will plot the function.
```
100 PRINT AT 6,0;"1"
110 PRINT AT 11,0;"0"
120 PRINT AT 16,0;"-1"
130 PRINT AT 21,1;"0"
140 PRINT AT 21,11;"2"
150 PRINT AT 21,21;"4"
160 PRINT AT 21,31;"6"
170 FOR X = 0 TO 6 STEP .1
180 PLOT X*10+2,INT(COS(X)10+20)
190 NEXT X
```

Note that the angle is in radians, not degrees. A radian is approximately 57 degrees. The precise conversion is:

180 degrees = PI radians.

CONT

This statement causes the program to continue running after it has stopped. When a program stops, a report is given to tell you why it stopped. The report will be of the form P/Q, where P is a letter or number, and Q is the line number.

If P is 9, then the program stopped at a STOP statement. If you use the CONT command, the program will restart at the line after the STOP statement.

example program.
100 PRINT "TYPE 1 TO CONTINUE"
110 INPUT A
120 IF A<>1 THEN STOP
130 PRINT "CONTINUING..."
140 GOTO 100

If you don't type 1, you will get the report 9/120. When you type CONT, the program will continue at line 130.

If P is not 9, then the program will continue at the line where it stopped. If the program stopped because of a programming error, this will probaply mean that it will stop again immediately. Usually this facility will be used after a BREAK.

COPY

This command does nothing if you haven't a printer attatched to your ZX81. If you have a printer, it will send a copy of whatever is on the screen to the printer. You can stop the printer at any time by using the BREAK key, and you will get report code D.

Sometimes, if you execute this statement without a printer attatched, the computer might get stuck. The break key will fix this problem.

DIM

There are two forms of DIM:- one for numeric arrays and another for string arrays.

1. DIM X(a,b,...z)

This will delete any array called X which already exists, and set up a new one. The letters a,b,..z have been used to represent numbers. They are the dimensions of the new array. For example DIM X(5,6) would set up a two dimensional array (a table) with 5 rows and 6 columns.

The new array will have all values set to zero. This feature could be used to zero an array in the middle of a   program.

An array cannot be used until it has been dimensioned in a DIM statement.

It is possible to have both an array and a variable called X. The computer will be able to tell which is which, because the array will have to be referenced using subscripts. It is not a very good idea though, because you might get confused when reading the program.

example program
```
100 DIM X(3,3)
110 FOR I=1 TO 3
120 FOR J=1 TO 3
130 LET X(I,J)=I*J
140 NEXT J
150 NEXT I
160 FOR I=1 TO 3
170 FOR J=1 TO 3
```

```
180 PRINT X(I,J);"  ";
190 NEXT J
200 PRINT
210 NEXT I
220 IF NOT X(1,1) THEN STOP
230 DIM X(3,3)
240 PRINT
250 GOTO 160
```

The first DIM statement is necessary before you can access the array. The effect of the second DIM statement is to zero the array.

2. DIM X$(a,b..z)

This works similarly to numeric arrays. Any string or string array will be deleted, so it is not possible to have both a string and a string array with the same name.

All strings in an array must be the same length. You can pad with blanks, if you want different length words in an array.

example program
```
100 DIM D$(7,6)
110 LET D$(1)="MON"
120 LET D$(2)="TUES"
130 LET D$(3)="WEDNES"
140 LET D$(4)="THURS"
150 LET D$(5)="FRI"
160 LET D$(6)="SATUR"
170 LET D$(7)="SUN"
180 INPUT N
190 PRINT D$(N);"DAY"
200 PAUSE 100
210 GOTO 180
```

You can see from this that the strings are automatically padded with blanks.

EDIT

The EDIT mode enables you to change lines without having to type the entire line again.

When you are typing in a program you will notice an inverse ⟩ sign between the line number and the text on one of the lines. This is the current line. You can use the up and down arrows (shifted 7 and 6) to change the current line.

If you now type EDIT (shifted 1), the current line will be reprinted below the listing of the program. (If you type EDIT and nothing happens, it is because you have run out of memory. Type CLS NEWLINE, then try again).

You can now move the cursor left and right in this line using the right and left arrows (shifted 8 and 5). The RUBOUT key will delete the character or key word immediately preceding the cursor. You can insert a character or key word at the position of the cursor by just typing it.

You are automatically in EDIT mode when typing in a new line. This means that you can use the edit commands (right and left arrows, rubout and insert at cursor position) to edit the line you are typing in.

When you have finished editing the line, type NEW LINE, and it will be inserted in the program in the correct position according to line number. If there is already a line with that line number, the old line will be deleted.

You can change the line number while you are editing. If you do, you will find that you have two copies of the line when you finish. When you ask to edit a line, a copy of the line is made; it is not automatically removed from the program.

EXP

The EXPonential function computes e**X where X is the argument given to the function. EXP always returns a positive number, since e = 2.71826 (and all powers of it are therefore positive). The EXP function gets large very quickly, so if you give it a number greater than 83 it will cause an arithmetic overflow.

This program will plot the EXP function

```
100 PRINT AT 15,0;"1"
110 PRINT AT 10,0;"2"
120 PRINT AT 21,1;"-4"
130 PRINT AT 21,11;"-2"
140 PRINT AT 21,21;"0"
150 PRINT AT 21,31;"2"
160 LET X=-4
170 FOR Y=2 TO 62
180 PLOT Y,INT(EXP(X)*5 +2)
190 LET X=X+.1
200 NEXT Y
```

FAST

Your ZX81 has two speeds of operation - fast and slow. When first switched on, the computer will be in slow mode. The FAST command changes it into fast mode.

In slow mode, the display file is constantly being written to the screen. Computations are done during the spaces between writing the display file.

In fast mode, the computer forgets about the screen except when it has nothing else to do; that is, during a PAUSE or while waiting for INPUT data from the keyboard. This means that the screen is blank while computations are being done, but the computer works about 4 times faster.

```
100 LET X=0
110 GOSUB 500
120 CLS
130 FAST
140 LET X=100
150 GOSUB 500
160 STOP
500 FOR I=X TO X+40
510 PRINT I,
520 NEXT I
530 RETURN
```

Note the difference between the times it takes to compute the numbers to be printed in fast and slow modes.

FOR...TO

A FOR statement must be used in conjunction with a NEXT statement later in the program.

```
100 FOR X=1 TO 10
110 PRINT X
120 NEXT X
```

In this program, the X is called the 'control variable'. If X had been previously used in the program, the previous X would be deleted.

When line 100 is executed, X is set to 1 (the first number specified). When time the program comes to the NEXT statement, it looks for a control variable called X. This variable is incremented by 1, and then tested to see if it is smaller then the limit (the second number specified). If it is larger than the limit, the program continues at the statement after the NEXT statement. Otherwise it jumps to the statement after the FOR statement.

An equivalent way of writing this program would be:-
```
100 LET X=1
110 PRINT X
120 LET X=X+1
130 IF X<=10 THEN GOTO 110
```

You can see from the program that the control variable may be used within the loop in the same way that any other variable.

If the program comes to a NEXT statement, and the variable name is wrong, there are two errors you may get. Error 1 means that the variable with the name you gave is not a control variable; it is an ordinary (simple) variable. Error 2 means that there is no variable with the name you gave.

FOR..TO..STEP

A FOR statement must be used in conjunction with a NEXT statement later in the program.

```
100 FOR X=10 TO 1 STEP -1
110 PRINT X
120 NEXT X
```

In this program, the X is called the 'control variable'. If X had been previously used in the program, the previous X would be deleted.

When line 100 is executed, X is set to 10 (the first number specified). When time the program comes to the NEXT statement, it looks for a control variable called X. This variable is incremented by the value of step (in this case -1), and then tested to see if it is between the start value and the limit (the second number specified). If it is not in this range, the program continues at the statement after the NEXT statement. Otherwise it jumps to the statement after the FOR statement.

If there is no step value, the computer assumes 1.

An equivalent way of writing this program would be:-
```
100 LET X=10
110 PRINT X
120 LET X=X-1
130 IF X>=1 THEN GOTO 110
```

You can see from the program that the control variable may be used within the loop in the same way that any other variable.

If the program comes to a NEXT statement, and the variable name is wrong, there are two errors you may get. Error 1 means that the variable with the name you gave is not a control variable; it is an ordinary (simple) variable. Error 2 means that there is no variable with the name you gave.

GOSUB

A GOSUB statement puts the line number of the next line in the program on the 'GOSUB stack', then jumps to the line specified in the statement. When a RETURN statement is encountered, the last line number to be put on the stack is taken off, and the computer executes that line.

example program
100 INPUT N
110 GOSUB 1000
120 PRINT N

1000 IF N<0 THEN RETURN
1010 LET N=SQR N
1020 RETURN

This program will calculate the square root of the number that is input. If the number is less than zero, the program will not change the value of N.

There are two RETURN statements in the program, but only one of them will be encountered.

Another way of writing this would be:-
100 INPUT N
110 GOTO 1000
120 PRINT N

1000 IF N<0 THEN GOTO 120
1010 LET N=SQR N
1020 GOTO 120

The advantage of using a subroutine is that it can be used in several places in tha program, and

it will always return to the line after the
calling statement.

A subroutine can be entered at any point. For
example, if you didn't want to test that N was
greater than zero, you could put:-

110 GOSUB 1010

The line number can be a variable or mathematical
expression too.

example program
100 INPUT N
110 GOSUB 1000*N
120 GOTO 100

1000 PRINT "SUBROUTINE 1"
1010 RETURN

2000 PRINT "SUBROUTINE 2"
2010 RETURN

3000 PRINT "SUBROUTINE 3"
3010 RETURN

This program will goto subroutine 1, 2 or 3,
depending on your input. If you input a number
greater than 3, it will stop with report code 0,
because there is no line number greater than
3010.

If your program tries to execute a RETURN
statement without having executed a GOSUB
statement, there will be no line number on the
stack for it to jump to. This will cause an
error, and the computer will stop execution with
report code 7.

GOTO

The GOTO statement causes the program execution to jump to the line number specified in the statement. This number can also be given by a variable or matematical expression. If the line does not exist, it will start execution at the next line with a greater line number than the one given. If the line number given is greater than any number in the program, execution will stop, with report code 0.

example program
100 INPUT N
110 GOTO N
120 PRINT "120"
130 PRINT "130"
150 PRINT "140"
160 GOTO 100

Try entering a few different numbers and see what happens.
If your number is less than 100, the computer will wait for another number to be input. If it is larger than 160, execution will stop.

If you input 110, you will keep the computer busy until you press the break key. That is called an infinite loop, because the program would just keep GOing TO 110 forever.

GRAPHIC

By typing GRAPHIC (shifted 9), you enter graphics mode, and a lot of different characters are available to you. You will notice that the cursor changes to an inverse G. Press GRAPHIC again, and you will be back into letter mode.

In graphics mode, all the numbers and letters are printed inverse (white on black). Symbols like +,- etc are also printed inverse by pressing the appropriate key with the shift key held down. Some keys have special graphics characters printed in the bottom left corner. These are obtained in graphics mode by pressing the shifted key.

example program
100 PRINT "▚▊"   (graphics E,E,5)
110 PRINT "▜▟"   (graphics R,R,1)
120 PRINT "▐▌"   (graphics 8,space)
130 PRINT "▙▊"   (graphics Q,8,4)

That will draw a picture of a man.

IF...THEN...

The IF statement enables the computer to make decisions. The format of the statement is:-

   IF condition THEN statement

If the condition is true (or non-zero) then the statement following the THEN will be executed. This statement can be anything at all; even another IF statement.

example program
100 INPUT AGE
110 IF AGE>21 THEN PRINT "YOU ARE AN ADULT"
120 PRINT "GOOD BYE"

The computer regards "true" and "non-zero" as the same thing. When it performs a comparison (like AGE>21) the ZX81 returns either a 1 (if the comparison holds) or a 0 (if the comparison does not hold). For this reason, there is no need to perform a comparison to see if a variable is zero.

IF N<>0 THEN statement

is exactly the same as

IF N THEN statement.

INKEY$

The INKEY$ function reads the keyboard to see if any key is being pressed. The result is the character being pressed if there is one key being pressed, otherwise it is the empty string (""). If there is no key being pressed, the computer does not wait; it keeps going regardless.

example program
```
100 LET T=0
110 LET A$=INKEY$
120 IF A$<>"" THEN GOTO 200
130 LET T=T+1
140 GOTO 110
200 SCROLL
210 PRINT AT 20,0;"TIME△";T,A$
220 GOTO 100
```

While you do not press any key, the computer is still busy; it is counting the time until you press the next key.

INPUT

The INPUT command causes the program to halt execution until you input the data on the keyboard. When you finish the input, press NEWLINE and the program will continue. If you are in fast mode, the display file will appear on the screen while waiting for input.

1. INPUT X
The program waits for a numeric input. You can input a number (such as 12) followed by a NEWLINE. Then 12 will be assigned to X.

If you input a letter (like Y), the computer will look for a variable Y. If there is one, X will be assigned the value of X. If there is not, you will get, error 2. That is, an undefined variable has been used.

If the first thing you input is STOP, the program will stop with report D.

example program
100 LET X=30
110 INPUT Y
120 PRINT "Y EQUALS△";Y
130 RUN

Try a few different inputs. For example, try X (a defined variable), and another letter that is not a variable name.

2. INPUT X$
This enables you to input a string variable to the program. When it is waiting for a string

input, the computer will put a ' " ' at the beginning of the string. Also, STOP will not stop the program as it did in the case of numeric input.

example program
100 FOR I=1 TO 20
110 INPUT X$
120 PRINT X$
130 NEXT I

INT

The INTeger function returns the integer part of the number that is its argument. It always rounds down. For example INT 2.9 = 2 , even though 3 is closer to 2.9.

```
example program
100 FOR I = 1 TO 20
110 LET X=RND*100
120 PRINT "INT (";X;") = ";INT X
130 NEXT I
```

LEN

This function must be given a string as its argument. It returns the length of the string.

For example:-
 The length of "" is 0. The length of "STRING" is 6.

example program
```
100 PRINT "ENTER A WORD"
110 INPUT A$
120 PRINT A$
130 PRINT "THAT WORD HAS△";LEN A$;"△CHARACTERS"
```

LET

The LET statement assigns a value to a variable. It operates in slightly different ways for numeric and string variables.

1. LET X=Y

Y can be a number, a variable, or an expression that evaluates to a number, involving both numbers and variables. A variable cannot be used in the program until it has been assigned a value in either a LET or an INPUT statement.

example program
```
100 LET X=2
110 LET Y=X
120 LET Z=Y+X
130 PRINT X,Y,Z
140 LET Z=2*(Z+X)+Y
150 PRINT Z
```

In line 140, Z is used in a statement assigning something to itself. That is alright, as long as Z has been defined previously.

2a. LET Z$=X$

After the assignment, Z$ will be the same length as X$. X$ may be of the form A$+B$. That will cause B$ to be added to the end of A$ (concatenated).

example program
```
100 LET Z$="ABCDEF"
110 LET Y$=Z$
120 PRINT Z$,Y$
130 LET Y$="DEF"
```

```
140 PRINT Y$
150 LET X$=Z$+Y$
160 PRINT X$
```

Note that in line 130, Y$ becomes 3 characters long, even though it was 6 characters long previously.

2b. LET A$(1 TO 3)=B$
In this case, if B$ is longer than 3 characters, only the first 3 will be used (truncation). If B$ is less than 3 characters long, spaces will be put in the remaining places.

example program
```
100 LET X$="1234567890"
110 LET X$(1 TO 3)="ABCDEF"
120 PRINT X$
140 LET X$(4 TO 9)="XYZ"
150 PRINT X$
```


2c. LET X$(3)=Z$
Now consider a dimensioned string array. Suppose X$ is a 3 by 5 array. Then if Z$ is less than 5 characters, X$(3) will be padded out with spaces to make 5 characters. If Z$ is more than 5 cahracters, it will be truncated to fit into X$(3)

example program
```
100 DIM X$(3,5)
110 LET X$(1)="JANUARY"
120 LET X$(2)="MAY"
130 LET X$(3)=X$(1)
140 FOR I=1 TO 3
150 PRINT X$(I);".."
160 NEXT I
```

LIST

The LIST command sends a listing of the program to the television screen, starting at the line number you specify.

    LIST line number

The line number is optional. If you don't give a line number, the listing will begin at the beginning of the program.

If your program is too long for the screen, it will fit as much as possible on the screen, then stop with report code 4 or 5.

After a list command, the current line is the line specified in the command.

Try typing in a few program lines and experiment with the LIST command.

LLIST

The LLIST command sends a listing of the program to the printer, starting at the line number you specify.

   LLIST line number

The line number is optional. If you don't give a line number, the listing will begin at the beginning of the program.

After a LLIST command, the current line is the line specified in the command.

If you haven't a printer attatched, this command should do nothing. However, the computer can get stuck. If this happens, pressing the BREAK key will fix things. You can stop the printer at anytime with the BREAK key. It will give report code D.

LOAD

The LOAD command is used to load programs that have been saved on cassette back into the computer.
If you know the name of the program, you can type

   LOAD "name"

and computer will look for a program called "name" on the tape.

If you can't remember what the name of the program, you can type

   LOAD ""

and the computer will load the next program on the tape into memory.

Next, start the cassette playing, and press NEWLINE. (Make sure the ear socket of the computer is connected to the ear socket of the cassette first!)

While the program is loading you will be able to see on the screen two patterns; one corresponding to the silence between programs, and the other corresponding to the program. (If you listen to the tape, you will be able to hear the difference too).

After a while the computer should stop, and report 0/0. If it doesn't, press the BREAK key. The most likely thing to have gone wrong is that the volume is wrong, so try again at a different volume level. It is more sensitive to the volume if you haven't given the name of the program.

LN

The LN function returns the natural logarithm of the number you give. That is the logarithm to the base e.

If LN X = A, then e**A = X.

Since e is poitive (approximately 2.7) X can never be negative. If you input a negative number, then you will get an error report, A.

example program
```
100 PRINT AT 6,0;"3"
110 PRINT AT 11,0;"2"
120 PRINT AT 16,0;"1"
130 PRINT AT 21,10;"10"
140 PRINT AT 21,20;"20"
150 PRINT AT 21,30;"30"
160 FOR X=.5 TO 30 STEP .5
170 PLOT X*2,INT(LN X *10)
180 NEXT X
```

This will plot the LN function.

LPRINT

The LPRINT statement is like a PRINT statement, but the printer is used instead of the television screen. You can use all the special commands (commas, semicolons, and tabs) just as in the PRINT statement. For a full explanation of these, see PRINT.

The text is not printed immediately. It is printed when a new line of text is started. So a line will be printed if it becomes full, or if the statement doesn't end with a comma or semicolon. If a TAB or comma needs a new line, the old line will be sent to the printer.

The way LPRINT works is to store whatever is to be printed in a buffer. The buffer is one line long, so whenever the new items won't fit in the buffer, the buffer is emptied by sending it to the printer.

The BREAK key will stop the LPRINT statement. If you haven't a printer attatched, LPRINT should have no effect, but if the computer gets stuck, the BREAK key will remedy this.

example program
100 LPRINT "1","2","3","4"
110 LPRINT
120 LPRINT "4";"5";"6";
130 PRINT "THAT LINE WILL NOT BE PRINTED YET"
140 PAUSE 100
150 PRINT "IT WILL BE PRINTED NOW"

NEW

This command clears all the program variables, and the program out of memory. You use this command just before entering a new program to get rid of the old program.

To see what it does, type in a few lines of code, and LIST them to make sure they are there. Now type the command NEW. If you try to list the program now, you will find nothing there.

NEXT

A NEXT statement must be used in conjunction with a FOR statement before it in the program.

```
100 FOR X=10 TO 1 STEP -1
110 PRINT X
120 NEXT X
```

In this program, the X is called the 'control variable'. If X had been previously used in the program, the previous X would be deleted.

When line 100 is executed, X is set to 10 (the first number specified). When time the program comes to the NEXT statement, it looks for a control variable called X. This variable is incremented by the value of step (in this case -1), and then tested to see if it is between the start value and the limit (the second number specified). If it is not in this range, the program continues at the statement after the NEXT statement. Otherwise it jumps to the statement after the FOR statement.

An equivalent way of writing this program would be:-
```
100 LET X=10
110 PRINT X
120 LET X=X-1
130 IF X>=1 THEN GOTO 110
```

You can see from the program that the control variable may be used within the loop in the same way that any other variable.

If the program comes to a NEXT statement, and the variable name is wrong, there are two errors you may get. Error 1 means that the variable with the name you gave is not a control variable; it is an ordinary (simple) variable. Error 2 means that there is no variable with the name you gave.

NOT

NOT can be used in two different ways:

1. It can be used in the condition of an IF..THEN statement, like this:-

    IF NOT A$=B$ THEN PRINT "NOT THE SAME"

This does exactly the same thing as

    IF A$<>B$ THEN PRINT "NOT THE SAME"

If the condition after the NOT is true, then the message won't be printed. The message will only be printed if the condition is false.

```
example program
100 INPUT A$
110 INPUT B$
120 IF NOT A$=B$ THEN GOTO 150
130 PRINT "THEY ARE THE SAME"
140 GOTO 100
150 PRINT "NOT THE SAME"
160 GOTO 100
```

2. NOT can be used as a numeric operator.

    LET X=NOT Y

is equivalent to the following

    LET X=0
    IF Y=0 THEN LET X=1

If Y is 0, then X is assigned 1, otherwise, X is

assigned 0. NOT treats Y as a logical value — TRUE (non-zero) or FALSE (zero) — and X is assigned the opposite logical value.

OR

There are two ways of using OR:-

1. It can be used in the condition of an IF...THEN statement. For instance,

    IF (it is raining) OR (there are dark clouds)
    THEN (take umbrella).

There are three cases when you should take your umbrella:-

1- if it is raining.
2- if there are dark clouds
3- if it is raining and there are dark clouds

That is the way OR works.

The only time that you need not take ake your umbrella is if both of the conditions are not true.

example program
```
100 INPUT X
110 IF X<0 OR X>255 THEN GOTO 140
120 PRINT X,CHR$(X)
130 GOTO 100
140 PRINT X;" IS OUTSIDE THE RANGE"
150 GOTO 100
```

In this example OR is used to check that X lies within the range 0 to 255.

2. OR can also be used a numeric operator.

        LET X=2 OR A

If A is 0, then X will be assigned the value 2.
Otherwise X will be assigned 1. This is
equivalent to

        LET X=1
        IF A=0 THEN LET X=2

The second number is treated as a logical value
by the OR. It has either a TRUE (non-zero) or
FALSE (zero) value.

example program
100 INPUT A
110 LET X=2 OR A
120 PRINT X
130 GOTO 100

You can try inputing different values of A and
check how the OR works.

PAUSE

The PAUSE statement stops the program for the number of frames specified. (One frame is 1/50 second). If you are in fast mode, the display file will be displayed during this time. If a key is pressed, the PAUSE will be cut short, and the program continue immediately.

For example, a PAUSE 50 statement will make the program wait for 1 second before continuing with the next statement.

If the number is more than 32767, it effectively means "PAUSE until a key is pressed". If you don't press a key, the program will wait forever!

example program
```
100 PRINT "HOW LONG DO YOU WANT TO PAUSE (IN SECONDS)?"
110 INPUT N
120 PAUSE N*50
130 PRINT "THAT WAS ";N;" SECONDS"
```

The next program will show how the 'PAUSE forever' works.

```
100 PRINT "PRESS A KEY WHEN YOU ARE READY"
110 PAUSE 35000
120 PRINT "HELLO. SO YOU ARE READY NOW"
```

PEEK

The PEEK function returns what is stored at a particular address in memory. You can PEEK at addresses in both the ROM (Read Only Memory) and the RAM (Random Access Memory).

The ROM is the list of instructions that enable the ZX81 to understand BASIC. They are written in machine code, so you will not be able to understand the instructions. But you can have a look at them if you like.

example program
```
100 FOR X=0 TO 20
110 PRINT PEEK X
120 NEXT X
```

That will PRINT the contents of the first 21 locations (or bytes) in the ROM.

The RAM is the part of memory that the program and all the program variables are stored in. The addresses are from 16384 to 17407. You can store values directly in the RAM using POKE and retrieve the value with PEEK.

example program
```
100 FOR X=17300 TO 17320
110 INPUT Y
120 POKE X,Y
130 NEXT X
140 FOR X=17300 TO 17320
150 PRINT PEEK X
160 NEXT X
```

When typing in numbers, make sure they are between 0 and 255. (See POKE).

PI(π)

The value of this function is 3.141592653. Ten digits of it are stored in the computer, but only 8 will be displayed. It requires no arguments.

example program
100 INPUT D
110 PRINT D*PI/180

This program will convert the number of degrees you input to radians.

Make sure you use the key marked π ; if you type in PI, the computer will think you mean a variable, and tell you it is undefined.

PLOT

The ZX81 divides up the screen in two ways. There are 21 by 31 character positions that you can PRINT to. Each of these poaitions is also divided into four 'pixels'. You can black in any of these pixels with the PLOT statement.

The pixels are numbered from 0,0 at the bottom left hand corner to 63,43 at the top right hand corner. Note that this numbering is different to the numbering of the PRINT postions, which start from 0,0 in the top left hand corner.

Check what happens when you print to a position, and then PLOT one of the four pixels in that position with this program.

```
100 PRINT AT 10,0;"0123"
110 PAUSE 50
120 PLOT 0,23
```

All of the character disappears, at that position, but neighbouring print positions are not affected.

You can use the UNPLOT statement to change a pixel from black to white again.

POKE

The POKE command stores the number you give it at a particular address.

    POKE 17300,100

will store 100 at location 17300.

There are two types of memory in the ZX81:- RAM (Random Access Memory) and ROM (Read Only Memory). What is in the ROM is unalterable.

The addresses of the RAM are 16384 to 17407. However you cannot POKE indiscriminantly. The computer uses some of these addresses as a 'working pad', to keep track of the state of the system. If you POKE some of these, the computer may crash.

example program
100 FOR X=17300 TO 17320
110 INPUT Y
120 POKE X,Y
130 NEXT X
140 FOR X=17300 TO 17320
150 PRINT PEEK X
160 NEXT X

You can input any number from -255 to 255, but if the number is negative, the computer will add 266 to it and store that number. A number outside this range will cause error B.

You will not get an error if you try to POKE into the ROM. It will just have no effect. A number outside the range 0 to 65535 will cause error B.

PRINT

The PRINT command is followed by a list of items to be written to the display file. The items in the list are separated by commas or semicolons. If you are in slow mode, the items will appear on the television screen immediately, otherwise they will appear next time the display file is written. (See FAST and SLOW).

The items will printed at the current print position. This is a location in the display file. The memory address of this location is stored in a specially reserved location in memory.

The items in the list can be:-

1. A numerical expression.
This will be evaluated, and the answer printed. If the number is negative, a minus sign (-) will precede the digits.

If the number is greater than $10**13$, or less than $-10**13$, it will be written in scientific notation. It will also be in scientific notation if it is between $-10**-5$ and $10**-5$. For example, $2 * 10**14$ would be written 2E+14.

The number will be printed without trailing zeros after the decimal point. For example 2.0000 will be written 2.

2. A string expression.
If there are any key word tokens in the string, they will be expanded out in full, including preceding and trailing blanks.

The '"' token (shifted Q) will be written as ".

Any unprintable characters will be printed ' ? '. If string includes any of the character codes that are unused, they will also print as ' ? '.

If the string does not fit on one line, it will be continued on the next line.

3 TAB x
This moves the print position to column x, if x is less than 32. If it is more than 32, then a multiple of 32 is subtracted from x until it is less than 32. x must be between 0 and 255, or program execution will stop, and you will get error B.

The effect of this is to print the next item at x, as long as TAB is followed by a semicolon.

4. Nothing
This is used when all you want to do is move the print position.

A SEMICOLON leaves the print postion where it is after printing the item, so the next item will be printed immediately after it.

A COMMA moves the print position the minimum number of columns to leave it in column 0 or 16. So if the last item finished at column 18, and was followed by a comma, the next item will begin at column 0 of the next line. If the item finished at column 15, so that the print position is column 16, a new line would be started. The comma moves the print position at least 1 place.

If the PRINT statement doesn't end with a comma

or semicolon, the print position will be left at column 0 of the next line. So the effect of the statement PRINT with no items is to move the print position down one line.

example program
```
100 PRINT "A AND B SEPARATED BY A SEMICOLON"
110 PRINT "A";"B"
120 PRINT
130 PRINT "A AND B SEPARATED BY A COMMA"
140 PRINT "A","B"
150 PRINT "TESTING THE TAB";
160 PRINT TAB 20;"TAB 20"
170 PRINT TAB 100;"TAB 100"
190 PRINT EXP 50,EXP -50
```

PRINT AT

The ZX81 divides the screen into 21 by 31 squares, any of which can contain a character. You can specify any one of these positions in a PRINT AT statement, and the PRINT items will be printed, starting at that position.

The PRINT items are the same as an ordinary PRINT statement, and semi colons and commas have the same effect too. (see PRINT).

example program.
100 FOR Y=0 TO 21
110 PRINT AT Y,Y;Y
120 NEXT Y

The position specified must be on the screen, or you will get error B. That is, the first number must be between 0 and 21, and the second must be between 0 and 31.
The two numbers speciying the position are separated by a comma. These must be followed by a semi colon.

RAND

The RAND command is used in conjunction with the RND function. It affects the 'randomness' of RND.

RND is not really a random number generator, but it is pretty good. It is called a 'pseudo-random' function.

RAND x
x is a number between 0 and 65535. Except for 0, by using the same x, you can force the random function to start at the same place each time. That means that RND will give the same sequence of numbers each time. So much for random numbers!

example program
```
100 FOR N=1 TO 5
110 FOR I=1 TO 3
120 RAND N
130 FOR J=1 TO 5
140 PRINT RND,
150 NEXT J
160 PRINT
170 PRINT
180 NEXT I
190 PAUSE 100
200 CLS
210 NEXT N
```

You can see that you get the same sequence of numbers each time.

RAND 0 or RAND (they mean the same thing) work differently. It causes RND to act more randomly than using a number. It will cause the RND function to start at a different place each time, depending on the number of frames that have been displayed on the television so far. That comes closer to being random than before.

```
100  FOR I=1 TO 3
110  RAND
120  FOR J=1 TO 5
130  PRINT RND,
140  NEXT J
150  PRINT
160  PRINT
170  NEXT I
```

REM

The REM statement can be followed by any sequence of characters except newline. It will have no effect on the running of your program, but will appear in the listing.

This can be useful when you are saving and loading programs on tape so that you remember what name you gave the program. You could put details of how the program works into a REM statement, so that when you come back to it some time later, you will be able to understand what the program does.

example program
```
100 REM PROGRAM TO PRINT THE NUMBERS FROM 1 TO 10
110 FOR X=1 TO 10
120 PRINT X
130 NEXT X
```

RETURN

A RETURN statement can only be executed if there has already been a corresponding GOSUB statement executed. Otherwise the computer will stop with report code 7.

A GOSUB statement puts the line number of the next line in the program on the 'GOSUB stack', then jumps to the line specified in the statement. When a RETURN statement is encountered, the last line number to be put on the stack is taken off, and the computer executes that line.

example program
100 INPUT N
110 GOSUB 1000
120 PRINT N

1000 IF N<0 THEN RETURN
1010 LET N=SQR N
1020 RETURN

This program will calculate the square root of the number that is input. If the number is less than zero, the program will not change the value of N.

There are two RETURN statements in the program, but only one of them will be encountered.

Another way of writing this would be:-
100 INPUT N
110 GOTO 1000
120 PRINT N

1000 IF N<0 THEN GOTO 120

```
1010 LET N=SQR N
1020 GOTO 120
```

The advantage of using a subroutine is that it can be used in several places in tha program, and it will always return to the line after the calling statement.

RND

The RND function returns a psuedo-random number between 0 and 1. It is 'psuedo-random', because it generates a definite sequence of numbers. Using the RAND statement, you can force the RND to use the same sequence of numbers again and again. (see RAND)

You can use the function as if it really were random though. The numbers appear to be random.

example program
100 LET N=INT(RND*6+1)
110 LET M=INT(RND*6+1)
120 PRINT M;"△△";N
130 PRINT "TOTAL=△";M+N

This program will simulate throwing two dice, and print the total.

RUN

The RUN statement deletes the variables left in memory by the previous program, and starts the current program.

You can follow RUN by a line number. This will cause the program to start execution at the line you specify. If you don't say which line to start at, the computer will assume that you mean the first line.

RUN does not clear the screen. You can use it in a program to start the program again; for instance at the end of a game to start the game again.

example program
100 PRINT "LINE 1"
110 RUN 150
150 PRINT "LINE 2"
160 RUN 100
You type RUN to start the program. Then the computer will just keep alternately printing line 1 and line 2. If instead you type RUN 150, it will start by printing line 2.

SAVE

The SAVE statement is used to record the program and its variables on cassette. You must give the program a name before saving it.

The procedure to save a program on cassette is as follows:-
1. Type SAVE "PROGRAM" , but don't press NEWLINE yet. The word between the quotes will be the name of the program when it is on cassette.

2. Connect the microphone socket of the cassette player to the microphone socket on the computer. Put the cassette in, positioned at a blank part of the tape. Start the cassette player recording.

3. Press NEWLINE. For a few seconds, the television screen will be grey. This is a lead in to the program. Then the television screen will be covered with black and white stripes. This is the program. After a while, the computer will report back with 0/0. It has finished now, and your program is saved on cassette. If you rewind the tape a little way, and play it back, you will hear a high pitched buzz, followed by a softer buzz. The loud buzz is the recording of your program, and the soft buzz is the lead in.

To load the program back into the computer, you use the LOAD statement. For an explanation of how to use this statement, look under LOAD.

SCROLL

The SCROLL statement scrolls the display file up one line. That is, every line is moved up one line. The first line is lost, and the new last line is completely empty.

The print position is moved to the beginning of the last line. (The print position is where the next item will be printed.)

example program
```
100 INPUT A$
110 PRINT A$
120 SCROLL
130 GOTO 100
```

SGN

The SGN function returns 1 if the number is positive, -1 if the number is negative, and 0 if the number is 0.

That is
```
    IF X<0 THEN LET SGN=-1
    IF X>0 THEN LET SGN=1
    IF X=0 THEN LET SGN=0
```

example program
```
100 INPUT X
110 PRINT AT 20,0;"SGN ";X;" IS ";SGN X
120 SCROLL
130 GOTO 100
```

SIN

If you give the SIN function an angle (in radians) it will calculate the ratio between the hypotenuse (H) and the side opposite the angle (Y) for a right angle triangle.

$$SIN\ A = (Y/H)$$

The number returned will be between -1 and 1, because the hypotenuse will be at least as long as the other side.

This program will plot the function.
```
100 PRINT AT 6,0;"1"
110 PRINT AT 11,0;"0"
120 PRINT AT 16,0;"-1"
130 PRINT AT 21,1;"0"
140 PRINT AT 21,11;"2"
150 PRINT AT 21,21;"4"
160 PRINT AT 21,31;"6"
170 FOR X = 0 TO 6 STEP .1
180 PLOT X*10+2,INT(SIN(X)10+20)
190 NEXT X
```

Note that the angle is in radians, not degrees. A radian is approximately 57 degrees. The precise conversion is
180 degrees = PI radians.

SLOW

Your ZX81 has two speeds of operation - fast and slow. When first switched on, the computer will be in slow mode. The FAST command changes it into fast mode. The SLOW command changes it back to slow mode

In slow mode, the display file is constantly being written to the screen. Computations are done during the spaces between writing the display file.

In fast mode, the computer forgets about the screen except when it has nothing else to do; that is, during a PAUSE or while waiting for INPUT data from the keyboard. This means that the screen is blank while computations are being done, but the computer works about 4 times faster.

```
100 LET X=0
110 GOSUB 500
120 CLS
130 FAST
140 LET X=100
150 GOSUB 500
160 SLOW
170 GOSUB 500
180 STOP
500 FOR I=X TO X+40
510 PRINT I,
520 NEXT I
530 RETURN
```

Note the difference between the time it takes to compute the numbers to be printed in fast and slow modes.

SQR

The SQR function, returns the squareroot of the number you give it. Taking the squareroot of a number is the opposite of squaring it (multiplying it by itself).

To work out a squareroot, for instance SQR 9, think of a number which, multiplied by itself gives 9. The answer of course is 3.

   SQR (A*A) = A

You cannot have a squareroot of a negative number, since if you multiply a negative number by itself, you still get a positive number. If you give the SQR function a negative number, you will get error B.

example program
```
100 PRINT AT 21,21;"40"
110 PRINT AT 21,11;"20"
120 PRINT AT 21,1;"0"
130 PRINT AT 15,0;"2"
140 PRINT AT 10,0;"4"
150 PRINT AT 5,0;"6"
160 FOR Y=0 TO 60
170 PLOT Y+2,(SQR Y)*5+2
180 NEXT Y
```

This program will plot the SQR function on the screen. There are really 2 answers for every number you give the SQR function; one postitive and the other negative. That is because 3*3 is 9, and so is -3*-3. The ZX81 will only give the positive answer.

STOP

The program will STOP executing, and report code 9 will be printed. If you then type CONT, the program will begin again at the next line.

example program
100 PRINT "TO STOP, TYPE S"
110 INPUT A$
120 IF A$<>"S" THEN GOTO 160
130 STOP
140 PRINT "THE PROGRAM STARTED AGAIN AT LINE 140"
150 GOTO 100
160 PRINT "THE PROGRAM DID NOT STOP"
170 GOTO 100

This program will enable you to see what happens when you STOP, and then press CONT.

STR$

The STR$ function is given a number and it returns the string of characters that would be displayed if the number was a PRINT item. For instance STR$ -4 = "-4".

That means that the two statements
```
    PRINT -4
```
and
```
    PRINT STR$ -4
```
will do exactly the same thing.

example program
```
100 LET A$=""
110 INPUT X
120 LET A$=A$+STR$ X
130 SCROLL
140 PRINT A$
150 GOTO 110
```

A$ will be a string containing all the numbers you input one after the other.

TAN

If you give the TAN function an angle (in radians) it will calculate the ratio between the side adjacent to the angle (X) and the opposite side (Y) for a right angle triangle.

$$TAN\ A = (Y/X)$$

This program will plot the function.
```
100 PRINT AT 6,0;"1"
110 PRINT AT 11,0;"0"
120 PRINT AT 16,0;"-1"
130 PRINT AT 21,1;"0"
140 PRINT AT 21,11;"2"
150 PRINT AT 21,21;"4"
160 PRINT AT 21,31;"6"
170 FOR X = 0 TO 6 STEP .1
180 IF ABS(TAN X)<2 THEN PLOT X*10+2,INT(COS(X)*10+20)
190 NEXT X
```

The resriction is necessary in line 180 because TAN X becomes infinite when X is an odd multiple of PI/2.

Note that the angle is in radians, not degrees. A radian is approximately 57 degrees. The precise conversion is

180 degrees = PI radians.

UNPLOT

The UNPLOT statement is used after a PLOT statement to blank out a 'pixel' after the PLOT function has blacked it in.

The screen is divided into 64 by 44 pixels, numbered from 0,0 in the bottom left hand corner to 63,43 in the top right hand corner. There are 4 pixels for every character position.

example program
100 INPUT Y
110 INPUT X
120 PLOT Y,X
130 PAUSE 50
140 UNPLOT Y,X
150 GOTO 100

Try inputting numbers for which there is no screen position, and see what happens. You should get an error message, B.

USR

This function is used to call machine code subroutines. You give the starting address of the subroutine as a parameter. The value returned is whatever was in the BC register pair when the subroutine completed.

To use this function, you will first need to enter a machine code routine. Chapter 20 shows you how to do this.

If the address you give is larger than 65535 or less than 0, you will get error B.

VAL

The VAL function is given a string as its argument, and returns the VALue of the string when evaluated as a numeric expression. For instance VAL "4+5" = 9.

The string to be evaluated may contain variables, but there are more rules about its use then:-

1. If you are using VAL in a larger arithmetic expression, then it must be the first item. For instance, 7 - VAL"X" must be rewritten, - VAL "X"+7.

2. If VAL is used as a co-ordinate in a PRINT AT, PLOT or UNPLOT statement, it must be the first co-ordinate. For instance, you cannot have PRINT AT 4,VAL "X".

Note that these rules apply only if there are variables in the VAL argument.

example program
100 INPUT A$
110 PRINT "VAL ";A$;" IS ";VAL A$
120 GOTO 100

You can type in numeric expessions, like 4 + SQR 10. Try typing this, before you RUN the program.

    LET X=5 (without a line number).
    GOTO 100

Now input X, and the number 5 will be printed. You can also use X in expressions now.

Also Available

Printed in Dunstable, United Kingdom